Trattato dell'Alcibra Amuchabile

(anonimo)

A Guided Translation

Gary Towsley
Olympia Nicodemi

Milne Library
Geneseo, NY

Published by Milne Library
SUNY Geneseo
1 College Cir
Geneseo, NY 14454

ISBN: 978-1-956862-27-0

The cover features the first and last pages of the *Trattato dell'Alcibra Amuchabile*, which is housed Biblioteca Riccardiana, Florence, Ricc. 2263. The photos are by the author, reproduced with permission of the Ministry of Culture of Florence.

Contents

Preface

The pre-modern algebra practiced in Western Europe in the centuries between Fibonacci (c.1200 CE) and Cardano (c.1550 CE) played a critical role in the evolution of algebra that culminated with the breakthroughs of the 16th century. But aside from Sigler's translation of Fibonacci's *Liber Abaci* [5], there are very few English translations of source material from this era. Our primary intention for this work is to fill that gap. We offer a translation of *Trattato dell'Alcibra Amuchabile*, (Anonimo), a pre-modern algebra text from the mid-fourteenth century, originally written in the Tuscan dialect of Italian. It is a set of rules or algorithms for solving algebraic equations which are followed by a large set of solved problems that use these rules. The manuscript is housed in the Biblioteca Riccardiana di Firenze and accessed as Codice 2263. For our translation, we worked from a transcription made by Annalise Simi [2] that maintained all the original spellings, word order, and mistakes found in the original. It gave us much needed legibility. We also consulted a facsimile of the manuscript provided by the Biblioteca Riccardiana.

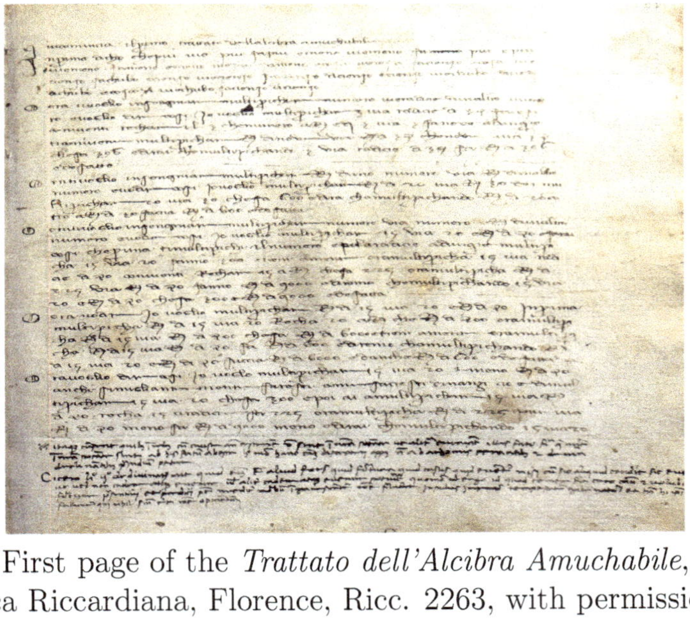

First page of the *Trattato dell'Alcibra Amuchabile*, Biblioteca Riccardiana, Florence, Ricc. 2263, with permission of the Ministry of Culture.

A look at the opening image tells us a story. Notice that there are no symbols, just words. This "rhetorical algebra" was the idiom of algebra from al-Khwarizmi to Cardano. It was inherited from a strictly oral tradition of teaching and learning where symbols were of no use. We keep this idiom in our translation. Where possible, we tried to keep the cadences of the original Italian in order to convey a way of thinking about algebra quite different from ours today. But it is not a word for word translation. Sometimes, the need for clarity took priority. We translated with the student of the history of mathematics in mind.

In addition to the actual translation of the *Trattato* our monograph provides ancillary material that situates the work historically, mathematically, and culturally. It provides a detailed guide to reading the algebra of 1350 and it provides pedagogical material in the form of suggested exercises. The actual mathematical skill set required of a reader does not exceed high school algebra. Our translation is a possible resource for a history course that is interdisciplinary.

This monograph is not a textbook. It presents material that can be sampled rather than swallowed whole. However, we suggest reading Chapters 1 and 2 before approaching the translation. Chapter 1 provides the historical context of the *Trattato.* Chapter 2 provides the tools to negotiate the mathematics: how it was written, taught, and thought of.

The *Trattato* was written about a century before the printing press was available. All works of the time were written by hand and many were subsequently copied by scribes. Ours seems to have been copied by a scribe (or scribes) for individual study, perhaps for or by a student. We refer to the writer (or writers) as "the scribe" or "the author" and use the pronouns "he/him" although nothing is known of the origins our tract.

Special thanks go to:
Weston Kennison, SUNY Geneseo
Jeffrey A. Oaks, University of Indianapolis
Librarians everywhere, especially Alana Nuth and Allison Brown, SUNY Geneseo

Chapter 1:
A Brief History of Pre-modern Algebra

The manuscript, *Trattato dell'Alcibra Amuchabile,* is an algebra text from the mid-fourteenth century, written in a Tuscan dialect of Italian. The author is unknown. Actually, almost everything about it is unknown. Before we look at the content of this manuscript we will place it in the history of the development of algebra. It has a very minor place in that development but it does tell us something about the big changes that were to come in the subject early in the sixteenth century.

The earliest appearance of written mathematics that might be called algebra seems to be in Mesopotamia around 2000 BCE. Many thousands of mathematical documents have been preserved from that period. The documents were written in wet clay and hardened in the sun. They are indestructible compared to papyrus or other forms of copying material. The teachers and students who wrote these documents were able to solve equations of the first and second degree, though in a manner we would find quite foreign to our mathematical sensibilities. Here is an example of a quadratic problem translated into very modern English:

> I have added the area and the side of my square. The result $\frac{3}{4}$. What are the area and the side? Take one-half of 1, the number of the side. $\frac{1}{2}$. Multiply it by itself. $\frac{1}{4}$. Add this to $\frac{3}{4}$ giving 1. Take its square root. 1. From this subtract the $\frac{1}{2}$. Result $\frac{1}{2}$ which is the side. The area is $\frac{1}{4}$.

We let x be the side of the unknown square. We are asked to solve $x^2 + x = \frac{3}{4}$. What follows the statement of the problem is an algorithm or a recipe for the solution. A translation more faithful to the original would show the geometry underlying the solution and would lead one to the phrase "completing the square." We will return to this problem a bit later.

The problems addressed in these early Mesopotamian clay tablets included such equations (in modern form) as $x^2 - ax = b$ and simultaneous equations of the form $x + y = a$, $xy = b$. These were all given as problems with particular numbers that served as models for solving similar problems. This form of mathematics appears to have flourished for about two centuries and then almost disappeared. Much of this material would reappear around 800 CE in the work of various Muslim scholars, most notably al-Khwarizmi. We do not yet know how it was transmitted to al-Khwarizmi.

One of the earliest Greek efforts in algebra comes in the work of Diophantus of Alexandria (c.100 CE). He wrote a large book called *Arithmetica.* It begins with his symbolic notation with which he expresses what we would call equations. His equations involved only polynomials with integer coefficients and he sought only positive rational solutions. All his problems reduced to either linear equations or to taking simple roots. His work differed from most Greek mathematics of his time in that it was problem oriented rather than presented as theoretical discourse. We will not look at his notation but focus on one of his problems, namely problem 8 from Book II:

> "*To divide a given square into two squares.*"

Diophantus was asking for two rational numbers whose squares had a sum equal to a given square. For his solution he gave an example. He chose the square to be 16. In terms of our algebraic notation his next step was to assume that the smaller squares were x^2 and $(2x - 4)^2$. Thus he had $16 = x^2 + (2x - 4)^2 = x^2 + 4x^2 - 16x + 16$. This easily reduces to $5x^2 = 16x$ or $x = \frac{16}{5}$ and so $2x - 4 = \frac{32}{5} - 4 = \frac{12}{5}$. He finds that the two squares he sought were $\frac{256}{25}$ and $\frac{144}{25}$. Their sum is indeed 16. The algorithm implied by his example produces a solution (in fact many different solutions) for any given initial square.[1] This particular problem is the one that led Fermat to his conjecture that carried the name "Fermat's Last Theorem."

Next we highlight the work of Muhammad ibn Musa al-Khwarizmi (780 – 850 CE). He was a Muslim scholar who studied and worked in the House of Wisdom in Baghdad. He wrote a book whose title in translation is *The Condensed Book on Calculation of al-Jabr and al-Muqabala* (circa 825). The term *al-jabr* from which our word algebra comes and the term *al-muqabala* refer to operations which reduce an

[1]See Exercise 1 in Suggested Exercises for details on his choice of terms as in $2x$–4.

algebraic equation to its simplest form. Notice that the work we have translated has as its title *Trattato dell'Alcibra Amuchabile.* It uses the Italian forms of those two words. Al-Khwarizmi begins by saying that there are three forms of numbers. We would call two of the forms unknowns, but he called them numbers. These forms were the square of an unknown, the unknown itself which he called the thing or the root, and a definite positive whole or rational number. For a given problem a relation between these three kinds of numbers would be determined and then, using the operations of *al-jabr* and *al-muqabala*, the relation would be reduced to one of six standard forms. They are as follows:

1. Squares equal to roots ($ax^2 = bx$)
2. Squares equal to numbers ($ax^2 = c$)
3. Roots equal to numbers ($bx = c$)
4. Squares and roots equal to numbers ($ax^2 + bx = c$)
5. Squares and numbers equal to roots ($ax^2 + c = bx$)
6. Roots and numbers equal to squares ($bx + c = ax^2$)

These six forms were not written as symbolic equations but as verbal expressions. Teaching at his time did not involve writing, only speaking and listening. Al-Khwarizmi's *Al-Jabr* was the first written algebra treatise, yet it seems that he expected it to be spoken. We mention that because this "rhetorical algebra"—algebra in words only—remained the protocol until the seventeenth century.

We look at one example from al-Khwarizmi's book, an example of the use of case 4.

> "One square, and ten roots of the same result in thirty-nine dirhems."

A dirhem was a small coin but here "dirhems" stands for the numbers. In this problem there are $1 + 10 + 39$ numbers of the three different types. We would write this as $x^2 + 10x = 39$. How does al-Khwarizmi solve this problem?

> The solution is this: you halve the number of the roots, which in the present instance results in 5. You multiply this by itself; the product is twenty-five. Add this to thirty-nine; the sum is sixty-four. Now take the root of this which is eight, and subtract from it half the number of roots, which is five. The remainder is three. This is the root of the square which you sought for; the square itself is nine.

While the numbers involved are different, the rule for the solution is exactly the same as for the example problem from the Babylonians mentioned above. Al-Khwarizmi was working fairly close geographically to where the earlier problem was posed and solved, but he was working nearly three thousand years later. We do not know how he found this form of solution. We do not know if it came directly from the Babylonian tradition or indirectly via other stops on the way.

Al-Khwarizmi presents many more problems that use his six rules. Several of the problems deal with inheritance and investment, but it is difficult to know just how practical these procedures actually were in daily life. In the three centuries after al-Khwarizmi, many other mathematicians in the Islamic world continued to develop algebra, though in form, it did not change much. Probably the greatest of those who followed al-Khwarizmi was *Umar ibn Ibrāhīm Nīshābūrī,* better known in the western world as Omar Khayyam (1048 – 1131). Omar studied cubic equations and found geometric solutions to all forms of the cubic using conic sections. We say "all forms" because all the equations had only positive coefficients so there were many different cubic equations. This mirrors the three different forms of quadratic equations found in al-Khwarizmi. Omar was disappointed, however, in not being able to find rules for the cubic equations in the form that al-Khwarizmi presented for quadratic equations.

How did the knowledge of this algebra get to medieval Europe where our author lived? There is no clear answer to this question. There are many possible paths and it is difficult to determine which was the actual one. For many years it was assumed that the knowledge of Arabic algebra came to Europe through the work of Leonardo of Pisa, better known as Fibonacci (1170 – 1240). Fibonacci's father had a position in Bugia, a city in present day Algeria. He took his son, Leonardo, with him on his travels and his son learned both the practical and theoretical mathematics available at the time. He traveled beyond Bugia around the Eastern Mediterranean where he expanded his knowledge of mathematics. When he returned to Italy he wrote a book called *Liber Abaci*, or Book of Computation. He wrote several other mathematical works but this is the one that dealt with both Hindu-Arabic numeration and algebra. Its first edition was published in 1202, followed by a second edition in 1228. In the last chapter of this work Fibonacci presents the algebra of al-Khwarizmi in abbreviated form. The bulk of the book deals with the base 10 place value system of numeration and applied problems based on the arithmetic that Fibonacci had learned traveling with his father. It contains full coverage of the algorithms for the four arithmetic operations and hundreds of business-oriented problems using this arithmetic. Fibonacci's book was written in Latin, the scholarly language of the day in Europe. However, the author of our *Trattato* wrote in the Italian vernacular of his region. Recently this fact (and others) has led to some questioning of the place of Fibonacci in the growth of algebra in Europe. There were several translations of al-Khwarizmi's algebra into Latin in the eleventh and twelfth centuries. These translations were made in Spain and Sicily, areas where Muslim and Christian scholars could interact and learn from each other. There is evidence that this algebra was translated into various vernacular languages, chiefly French, Italian, and Catalan. This may have been the route by which our author obtained his knowledge of algebra. Regardless of the actual route which algebra took to get to medieval Italy, we now look at how algebra was used in medieval Italy and by whom.

In the period 1200 – 1600, the larger cities in Northern Italy had "grammar" schools which taught reading, writing, and grammar to students, mostly boys, of ages from seven to eleven. The instruction was both in Latin and in the local dialect of Italian. During this period there were also "abbacus" schools that offered a two-year program to eleven to thirteen year-olds as a follow up to grammar school. The word "abbacus" does not refer to the calculating device that was used in its many forms throughout Asia, Europe, and Africa. Instead, it refers to calculation. Hence these were schools of calculation or computation. The curriculum in these schools trained students in arithmetic using Hindu-Arabic numeration which was then applied to problems with a commercial setting. These schools provided the booming trade and financial centers of Northern Italy with a numerate workforce and contributed greatly to the growing wealth of the region. Some of the schools were privately run while others were run by the cities. The curriculum for these schools was spelled out in books called "abbacus books" often written by those who taught in these schools. The treatise we have translated is one of those abbacus books. However, there is a large problem with that designation. Algebra was never, or at most rarely, taught in the abbacus schools. Our treatise is almost entirely devoted to algebra. Why was this book written and who might its audience have been? There are several possible answers. First, the potential teachers of the abbacus schools needed to convince the schools that they were worthy of being hired to teach commercial arithmetic to these selected students. One way to gain the confidence of the hiring committee was to write a book that went beyond the standard curriculum. To write a book on algebra that was full of numerical problems, the applicant certainly must have been proficient in the more basic arithmetic. Second, there are some records from northern Italian cities of an interest in algebra as a problem-solving tool for architects, surveyors, engineers, and others in related fields. One important fact to note is that during the Middle Ages algebra was not part of the mathematics studied in the universities. There the curriculum in mathematics consisted of the "quadrivium," the name given to the four portions of the seven liberal arts devoted to mathematics. The quadrivium consisted of arithmetic (number theory), geometry, music, and astronomy. It was studied in the universities through Greek texts that were for the most part translated into Latin. Algebra was not found in the quadrivium. In growing cities like Florence, Siena, Venice, and Milan, there was a significant population that wanted to learn mathematics that would be useful to them but did not want to study it in the universities. Books such as our treatise would have found a ready audience there.

How did our treatise from the middle of the fourteenth century compare with the work of al-Khwarizmi? The *Trattato* contains all the rules of al-Khwarizmi's work and then adds some extensions. Cubic and quartic equations that can be reduced to quadratics are solved correctly. Cubic equations that do not reduce to quadratics were solved incorrectly. The rules for these cubics were simply the rules for the quadratics most similar in form to them. Our author was not alone in presenting false rules for cubic equations. Many other authors of abbacus books, both before and after ours, contained the same errors. The great bulk of the treatise consists of problems. Al-Khwarizmi's text is similar in that way. The commercial problems however, were from a different society with different concerns. One feature of the algebra of our text that was missing from al-Khwarizmi was the use of algebraic fractions, that is expressions in which the divisor contained the unknown or*cosa*, the Italian term for the unknown.

There were many abbacus books written in the period 1200 – 1500, differing in content and style but not going far beyond what was laid out in al-Khwarizmi's work. One exception is found in the work of Master Dardi (from Pisa in the fourteenth century) who was able to solve certain cubic equations, ones in which there was a specific relation between the terms. He could not solve a general cubic. In 1494 Luca Pacioli (1445-1517), a friend of Leonardo da Vinci and of the artist Piero della Francesca, published what was then an encyclopedia of mathematics called *Summa de arithmetica, geometria, proportioni et proportionalita.* He noted in his work that there did not exist an algebraic solution of a general cubic equation, noting the tradition of incorrect solutions found in abbacus books.

Pacioli's book was written in Italian and was consulted by university mathematicians. By 1500 the algebra found in the abbacus books had come to the notice of university mathematicians. This was partly due to the inclusion of problems requiring algebra for their solution in the contests held between mathematicians. These contests offered prizes to the winner and were a very positive part of the resume of someone seeking to teach mathematics at the university level. Sometime around 1515 a professor at the University of Bologna, Scipione del Ferro (1465 – 1526), solved one form of the cubic equation algebraically. The form he solved would have been described as *Cubes and Things equal Numbers.* In modern notation he had solved $x^3 + px = q$ where p and q are positive numbers. He did not publish or even announce his breakthrough. He kept it private for his use in mathematical contests. He could pose problems that no one else could solve quickly. Before he died del Ferro gave his secret to one of his students, Antonio Maria Fiore, and to a colleague at Bologna, Annibale della Nave. Fiore challenged another mathematician, Nicolo Tartaglia, (1499 – 1557) to a mathematical contest. Spurred on by this contest Tartaglia discovered the solution to a different form of the cubic, namely cubes and squares equal numbers ($x^3 + px^2 = q$) as well as to the form that del Ferro had discovered. Tartaglia won the contest. Because the final forms of the cubics that were to be solved had only positive coefficients there were 13 different forms of cubic equations. Only two had been solved by the time of the Fiore-Tartaglia contest. We will not go into the story of how the solutions that Tartaglia knew leaked out but indeed they did. Their recipient was one Girolamo Cardano (1501 – 1576).

With the help of his student and colleague, Ludovico Ferrari (1522-1565), Girolamo Cardano was able to solve all the forms of the cubic equation and to do the same for fourth degree or quartic equations. He published all this in a book titled *Ars Magna.* In this work he gave the solutions to the different forms of the cubic as algorithms or recipes by using particular numerical examples. For instance, in Chapter 11 he solves *"Cubes and things equal to numbers."* But his solution is for the specific case of *a cube and six things equal to twenty.* He did not give a formula that would work for all cubics of that form, but instead, gave a procedure to follow. There were several advances documented in this work. First, Cardano allowed the equations he solved to have negative roots. He called them false or fictitious roots. Second, he shows how to alter a cubic equation that contains a square term to an equivalent one that does not have such a term. He had already given rules for the solutions to all these. Finally, he solves a quadratic equation that we have already seen but with a twist. The equation was of the form *squares and numbers equal to things.* The particular one he solved was *a square and 40 numbers equals 10 things,* ($x^2 + 40 = 10x$). The roots are $x = 5 \pm \sqrt{-15}$, complex numbers. After solving this problem he wrote "So progresses arithmetic subtlety, the end of which, as is said, is as refined as it is useless."

In his book, *Algebra,* another Italian mathematician, Rafaello Bombelli (1526 – 1572) presented the material from Cardano's *Ars Magna* in a more organized and readable fashion. He provided a guide to the arithmetic of complex numbers because this arithmetic was very helpful in solving cubic and quartic equations. He also expanded the use of symbols in his algebra.

The next big step in algebra was taken by Francois Viete (1540 – 1603). In a series of papers or treatises Viete transformed algebra to something like its modern form. His biggest advance was the introduction of unknown but fixed coefficients for terms in an algebraic equation. Instead of dealing with particular quadratic equations such as $x^2+10 = 39$, he was able to work with expressions like $ax^2+c = bx$ where a, b, and c are positive but unspecified numbers. He used vowels for the variables and consonants for the constants. His quadratic equation might look like $A^2 + C = BA$. His actual expressions were somewhat more complicated but we will not to go into those details. Whereas Cardano had given rules by which a root of a particular cubic equation could be found, Viete was able to use a substitution and derive a formula for the root in terms of general coefficients. With Pierre de Fermat (1601 – 1665) and Rene Descartes (1596 – 1650) we finally find a recognizable algebra of polynomials in one or more variables with the relation between roots and factors made explicit. They both claimed Viete as the inspiration for their works.

So where does this place our treatise in the development[2] of algebra? It showed a small advance on the algebra of al-Khwarizmi in dealing with higher powers of the unknown (not always correctly) and in working with algebraic fractions. Our treatise was written in the Tuscan dialect of Italian and not in the Latin of the mathematics of the University and was aimed at interested and educated readers who were not necessarily connected to the universities. Numerous works like it sustained the discipline for three centuries in Europe until it was ready for the sixteenth century breakthroughs.

[2] *A History of Mathematics* by Victor J. Katz, number 13 in References and Resources, is a deep and accessible source of information about the development of mathematics. Original source material is found in numbers 1, 2, 3, 5, 8, 10, and 16.

Chapter 2: Selected Problems Annotated

The *Trattato* comes in three sections. The first deals primarily with the rules for the arithmetic of radicals, the second with 24 rules of algebra (algorithms, actually), and the third presents problems that use these rules. In this chapter we will present several fully annotated examples that will introduce the reader to the vocabulary and methodology of our *Trattato.* They are typical of pre-modern algebra manuscripts in the abbacus tradition. But first, we ask the reader to take a quick look at the translation of the *Trattato* in Chapter 3. Notice that there are no symbols, no x's, $+$ or $-$ signs, or radical signs. In the original manuscript, there was also no punctuation. It is all in words. We call this rhetorical algebra. Algebra and arithmetic treatises were written this way from the time of al-Khwarizmi in the 9th century CE until the 16th century.

Note: In quoted texts, a number in brackets that follows a phrase or sentence indicates that an annotation with the same number follows below.

Example 1.
We start with the very first paragraph of the manuscript.

> First I say that *added* by *added* makes *added* and *subtracted* by *subtracted* makes *added* and *added* by *subtracted* makes *subtracted* and *subtracted* by *added* makes *subtracted* [1] and *thing* by *thing* makes *censo* and *thing* by *censo* makes *cube* and *censo* by *censo* makes *censo di censo* and *censo* by *cube* makes *censo di cube* and *thing* by *cube* makes *censo di censo.*[2]

[1] The sentence starts with what looks like a rule for multiplying signed numbers, "First, I say that added by added makes added subtracted by added makes subtracted ..." But it is not. Numbers were all strictly positive; there were no negative numbers. But there was subtraction. The instructions provide a guide for multiplying binomials with added and subtracted terms. For example, each of the given cases is required to expand $(2 - \sqrt{3}) \cdot (4 - \sqrt{5})$ correctly.

[2] Next we have rules for multiplying the unknowns. In rhetorical algebra the unknown, what is usually x for us, is called "a *thing*." It is *cosa* in Italian, *shay* in Arabic, and *res* in Latin, all words that mean "thing." The goal of an algebraic problem is to find the value of the *thing*. The product of a "*thing* by a *thing*" is called a *censo* (pl. *censi*) in Italian, x^2 to us. It is called *mal* in Arabic and *censum* in Latin. All these words refer to property or money. We keep this historical usage. *Censo* by *thing* makes a *cube* (*cubo* in Italian, x^3 to us); *censo* by *censo* makes a *censo di censo* (x^4); *censo* by *cube* makes *censo di cube* (x^5); *thing* by *cube* makes *censo di censo* (x^4). Numerical exponents did not enter algebra until much later.

The first sentence about multiplication rules is followed by 14 rules for computations involving roots. We will look at the first and the eighth of these rules.

Example 2.
The first rule about roots tells us how to multiply a number by the root of another number.

> Now I want to teach you to multiply a number by the root of another number and I want to say it this way: I want to multiply 3 by the root of 34. Do it this way: you must bring the 3, which is a number, to the root and say 3 by 3 makes nine. So you must multiply the root of nine by the root of 34, which is nine by 34 which makes 306 and say that multiplying 3 by the root of 34 makes the root of 306 and it is done.

The author works with square roots and cube roots throughout the treatise. It was important to be able to compare two numbers and determine which was greater. For instance, the difference of two quantities was always computed as the smaller from the larger, necessarily positive. Which is greater, $3 \times \sqrt{34}$ or $2 \times \sqrt{75}$? It is not obvious.

[1] To multiply $3 \times \sqrt{34}$, the author "brings the 3 which is a number to the root." We would say "Bring 3 under the radical sign." To do so he must multiply 3 by 3 and then multiply 9 by 34. The answer: 3 by the root of 34 is the root of 306 i.e. $3 \times \sqrt{34} = \sqrt{306}$. Is it larger

or smaller than $2 \times \sqrt{75}$? Larger since $2 \times \sqrt{75} = \sqrt{4 \cdot 75} = \sqrt{300}$.

Example 3.
The eighth rule shows how to multiply two binomials. It uses the instructions for added and subtracted quantities from the first paragraph. It also contains a diagram in the form of a cross or large X to guide the multiplication.

> Now I want to say: I want to multiply 5 and the root of 20 by 5 less the root of 20.
>
> 5 plus $\mathfrak{R}$ of 20
> via
> 5 less $\mathfrak{R}$ of 20
>
> Again similarly, you place it as you placed the one before [1] and then multiply also as you did the one before, that is, multiply 5 by 5 to make 25, [2] and then multiply by the cross: 5 by the subtracted root of 20. [3] You must bring 5 to the root, which is 25. Now multiply the root of 25 by the subtracted root of 20. It makes the subtracted root of 500. [4] Now multiply the other the other piece of the cross, that is 5 by the added the root of 20 which is the added root of 500. Join it with 25 and the subtracted root of 500. They make exactly 25. [5] Now it remains for you to multiply the added root of 20 and the subtracted root of 20 which is a subtracted 20. Between it and 25, there remains 5 and we will say that, multiplying 5 and the root of 20 by 5 [less] the root of 20 which makes 5 exactly and it is done. [6]

[1] There was a similar diagram in the previous problem. In the original manuscript, this problem has a little diagram scribbled off to the side. The reference to "cross" that follows suggests that the scribe's source had a diagram that looked more like:

5 plus $\mathfrak{R}$ of 20
X
5 less $\mathfrak{R}$ of 20

[2] First multiply 5 by 5 to get 25.

[3] Then he computes 5 times the subtracted root of 20, the terms on the negatively sloped diagonal.

[4] He "brings it to the root" to get the subtracted root of 500. Subtracted from what? From 25. So far he has $25 - \sqrt{500}$.

[5] Similarly, he computes via the other diagonal, which is 5 times the added root of 20, to get $\sqrt{500}$. He then adds that to $25 - \sqrt{500}$ to get 25.

[6] Finally he computes root of 20 by the subtracted root of 20 to get a subtracted 20. It is subtracted from 25. The result is "exactly 5."

The first six of the 24 computation rules for algebra are standard in most pre-modern algebra manuscripts. They are the six rules given by al-Khwarizmi in his *Al Jabr.* The first three rules deal with the simple cases $ax = b$, $ax^2 = b$, and $ax^2 = bx$. The last three deal with trinomial quadratic equations: $ax^2 + bx = c$, $ax^2 + c = bx$, and $ax^2 = bx + c$. Why three? Coefficients and solutions must all be strictly positive. The 22 the rules of the 24 listed in our manuscript that are correct all reduce to these six equations. We will look closely Rule 5, the instructions for solving the case $ax^2 + c = bx$. It is the only rule for a quadratic for which two solutions are presented. The instructions for Rules 4 and 6 are very similar.

Example 4.

> **Rule 5.** When the things are equal to censi and a number [1], you want to divide by the *censi*[2], then halve the *things* and multiply it by itself [3] and subtract the number [4]. The root of that which remains plus the half of the things is the value of the thing [5] or, in truth, the half of the things less the root of what remains. [6]

[1] Rule 5 gives instructions for quadratics of the form "things equal to censi and a number" or $ax^2 + c = bx$.

[2] The first step is to divide by the *censi* which is to divide all terms by the *number* of *censi*, or by what we would call its coefficient. The equation becomes $x^2 + \frac{c}{a} = \frac{b}{a}x$. The number of *things* is $\frac{b}{a}$ and the number is $\frac{c}{a}$.

[3] Next take half of the number of things and square it to get $\left(\frac{b}{2a}\right)^2$.

[4] Now we subtract $\frac{c}{a}$ from $\left(\frac{b}{2a}\right)^2$. What remains is $(\frac{b}{2a})^2 - \frac{c}{a}$.

[5] Then take the root of what remains and then add the number to get the value of the *thing*: $x = \sqrt{(\frac{b}{2a})^2 - \frac{c}{a}} + \frac{b}{2a}$.

[6] The other solution is, "in truth" valid as well: $x = \frac{b}{2a} - \sqrt{(\frac{b}{2a})^2 - \frac{c}{a}}$. Note that the radical is smaller than $\frac{b}{2a}$ and so this solution is positive.

The first task in solving any polynomial equation was to get it into the format addressed by a rule. For instance, if a problem results in an expression such as "3 things and 2 censi are equal to 5 censi plus 7 numbers less 6 things, ($3x + 2x^2 = 5x^2 + 7 - 6x$), then these 6 things (or $6x$) are missing from $5x^2 + 7$ and must first be restored by adding $6x$ to both sides. We get $9x + 2x^2 = 5x^2 + 7$. The expression must then be balanced by gathering like terms on the appropriate side of the equation. Our equation would be transformed into $9x = 3x^2 + 7$ making it ready for Rule 5, "When things are equal to censi and a number." The Arabic words for restore and balance are in the title of our manuscript. Restoring (*al-jabr*) is adding back what has been subtracted (removed). *Al-muq-abala* is to gather like terms on one side of an equation. The Arabic word means confrontation, "with the same face." So before going on to solving an equation in our *Alcibra Amuchabile*, all subtracted terms must be restored and all like terms gathered.

The next examples are two problems that follow Rule 5 to illustrate its use. Both are purely algebraic. The first problem is written out in full here and then parsed, step by step. The format is followed by almost every problem in the text. It may seem repetitive and formulaic but it comes from a time when instruction was oral and students had to "keep in mind" what was said. The second problem gives us insight into the mathematical mindset of the abbacus era.

Example 5.

> Make for me from ten two parts such that, multiplying the larger against the smaller, makes 20. I ask how much each part is. Suppose that the smaller is a thing. Therefore, the larger is what remains up to 10 which comes to 10 less a thing. Next you must multiply the smaller, which is a thing, by the larger, which is 10 less a thing, making 10 things less a censo, the multiplication of which is equal to 20. Restore each part which is to join a censo to each part. You will have that 10 things are equal to a censo and 20 numbers. Bring [it] to one censo and then take half the things, it comes to 5, and multiply it by itself, making 25. Subtract the number which is 20. There remains 5 and take the root of this, which clearly does not have roots. Therefore the value of the thing is that which was halved less the root of 5. And we had supposed that one part was the thing, the smaller one, thus the smaller is 5 less the root of 5 and the larger is what remains of 10 which comes to 5 plus the root of 5.

Problems that start with "Make for me from ten two parts" are ubiquitous in the abbacus algebra manuscripts and appear in the *Liber Abaci* of Fibonnaci (c. 1228, in Latin) and in the Arabic sources such as the work *Al-Jabr* of al-Khwarizmi (c. 800 CE). It is remarkable how many different problems can be initiated this way. What follows is the step-by-step breakdown of the problem stated above. The original problem is in italics.

a. *Make for me from 10 two parts such that, multiplying the larger against the smaller, makes 20. I ask how much each part is.* The problem is posed and what must be solved for is stated.

b. *Suppose that the smaller is a thing.* The unknown is identified.

c. *Therefore, the larger is what remains up to 10 which comes to 10 less a thing. Next you must multiply the smaller, which is a thing, by the larger, which is 10 less a thing, making 10 things less a censo, the multiplication of which is equal to 20.* The equation to be solved is established. We would say: "Suppose that x is the smaller thing, so the larger is $10–x$, and their product is $10x–x^2$ which must equal 20." The equation to be solved is thus formulated: $10x–x^2 = 20$.

d. The Algebra:

   i. *Restore each part which is to join a censo to each part. You will have that 10 things are equal to a censo and 20 numbers.* We "restore" by adding back the subtracted x^2 to each side to obtain $10x = x^2 + 20$. Now it is ready for the appropriate rule, in this case, Rule 5.

   ii. *Bring [it] to one censo* means to divide by the number of censi (even if the number is 1) as per Rule 5 instructions.

   iii. *Then take half the thing, it comes to 5, and multiply it by itself, making 25. Subtract the number which is 20. There remains 5 and take the root of this, which clearly does not have roots. Therefore the value of the thing is that which was halved less the root of 5.*
   As the rule says, we take half of 10 (things) which is 5 and square it to get 25. Subtract 20 from 25 to get 5 (which not a perfect square) and conclude that our unknown x is equal to $5–\sqrt{5}$. The author opted to subtract the radical since the unknown is the smaller quantity and the alternative choice would have made it the larger.

e. *And we had supposed that one part was the thing, the smaller one, thus the smaller is 5 less the root of 5. And the bigger is what remains of 10 which comes to 5 plus the root of 5. The smaller is* $5–\sqrt{5}$ *so the larger is* $10–(5–\sqrt{5})$ *or* $5 + \sqrt{5}$. The author summarizes and brings us back to the original problem.

In the middle of Rule 5, the author takes a moment to praise his rule but also gives a caveat.

> Made this way, this rule is much to be praised because it gives two answers and you are as good with one as the other. But we must keep in mind that not all problems that use this rule can be answered with two answers if one does not check them.

He then gives a problem that, for him, proves that sometimes, one of the solutions is incorrect. It is another problem that starts, "Make for me of 10 two parts such that."

Example 6.

> Make for me of 10 two parts such that, multiplying one against the other, and joined onto this multiplication the difference of one part from the other, makes 22. I want to know how much each part is. Suppose that one part is a thing and thus the other is 10 less a thing. Now multiply one against the other, which is one thing by 10 less one thing, making 10 things less a censo. Now, after this multiplication, you want to join the difference which is one thing from 10 less a thing, making 10 less 2 things. To this difference, you want to join 10 things less a censo making 10 and 8 things less a censo which is equal to 22.

The parts of 10 are x and $10–x$ and their product is $10x–x^2$. He takes the difference to be $(10–x)–x = 10–2x$. The equation to solve is thus $10+8x–x^2 = 22$. He proceeds to the algebra, restoring and gathering, to obtain: $x^2 + 12 = 8x$, ready for Rule 5.

The Rule gives us two solutions, $x = 6$ and $x = 2$. We check by substituting into $x^2 + 12 = 8x$ and find them both to be correct. But he rejects $x = 6$. (Note that if $x = 6$, the other part of 10 is 4.) He says:

> Multiply one against the other to make 24. Add the difference which gives 4 from 6 which is 2, making 26 which was supposed to be 22. Therefore we can see clearly that the value of the thing is not the radical of what remains plus half of the thing. **So we have proved that is not true.**

The problematic phrase is " the difference which gives 4 from 6 which is 2," always smaller from larger. But in the setup of the equation, the difference of the parts of 10 is $(10–x)–x = 10–2x$. With $x = 6$, the difference is negative. But notice how the author tested 6. He did not substitute it into his equation, $x^2 + 12 = 8x$. He went back to the narrative of the problem: "Make for me from 10 two parts which, multiplied one against the other and onto this multiplication is joined the difference of one part from the other ..." With $x = 6$ and $10 - x = 4$, this becomes, "Multiply one against the other to make 24. Add the difference (always positive) which gives 4 from 6 which is 2, making 26 which was supposed to be 22."

In pre-modern algebra, the unknown *thing* (*cosa*) was not a variable into which you could substitute arbitrary values. It had a value, unknown to start, known after it was solved for. The correctness of its value did not depend on substituting into the equation that resulted from the words, but from the words themselves. Polynomials were not yet functions in 1365. They were simply inventories of terms, and the inventory directs how the rule is applied. Beyond that, the polynomial loses its connection to the problem. In short, we find a very different mindset in the algebra of 1365.

Rules 7 through 11 give instructions for solving cubic equations that reduce to linear or quadratic equations. For example, Rule 10 addresses "When the cubes are equal to censi and to things" or $ax^3 = bx^2 + cx$. We would posit the solution $x = 0$ (not an option here), then cancel, and proceed solve $ax^2 = bx + c$ according to Rule 6. Here, there is no such reduction, only instructions in terms of the coefficients. That process goes awry for Rules 12 and 13.

Example 7.

> **Rule 12.** When the cubes are equal to things and a number, [1] you must divide by the cubed things, then halve the things and multiply it by itself and put it with the number and the root of that plus half of the things is the value of the thing.[2] Find for me two number such that the parts, one to another, are as 2 to 3 and multiplying the first by itself and then by that number, makes as much as the other plus 12. [3]

[1] The equation to solve is $ax^3 = bx + c$

[2] The instructions are exactly those of rule 6 that solve $ax^2 = bx + c$

[3] The example that illustrates the rule leads to the equation $8x^3 = 3x + 12$. The rule as stated is correctly applied and gives the incorrect solution to the cubic equation $x = \sqrt{1\frac{274}{512}} + \frac{3}{2}$. The answer is not checked by substitution, not surprising (see Examples 5 and 6), but also not checked against the words of the original problem, perhaps more surprising.

Chapter 3

Trattato dell'Alcibra Amuchabile

(an English Translation)

The first treatise on Alcibra Amuchabile begins.

1. First I say that added by added makes added and subtracted by subtracted makes added and added by subtracted makes subtracted and subtracted by added makes subtracted and thing by thing makes censo and thing by censo makes cube and censo by censo makes censo di censo and censo by cube makes censo di cube and thing by cube makes censo di censo.

2. Now I want to teach you to multiply a number by the root of another number and I want to say it this way: I want to multiply 3 by the root of 34. Do it this way: you must bring the 3, which is a number, to the root and say 3 by 3 makes nine. So you must multiply the root of nine by the root of 34, which is nine by 34 which makes 306 and say that multiplying 3 by the root of 34 makes the root of 306 and it is done.

3. Now I want to teach you to multiply the root of a number by the root of another number and I want to say it this way: I want to multiply the root of 20 by the root of 30. You must multiply 20 by 30 which makes 600 and say that multiplying the root of 20 against the root of 30 makes the root of 600 and it is done.

4. Now I want to teach you to multiply a number by a number and the root of another number and I want to say it this way: I want to multiply 15 by 20 and the root of 40 and you will do this way, that first you multiply the number and then the root. So multiply 15 by 20 making 300 and keep it in mind. Now multiply 15 by the root of 40. You must bring 15 to the root which makes 225. Now multiply the root of 225 by the root of 40 making the root of 9000 and we say that, multiplying 15 by 20 and the root of 40 makes 300 and the root of 9000 and it is done.

5. Now I want to say: I want to multiply the root of 15 by 20 and the root of 40. First multiply the root of 15 by 20. Bring 20 to the root which is the root of 400. Now multiply the root of 15 by the root of 400, which makes the root of 6000 and keep it in mind. Now multiply the root of 15 by the root of 40, making the root of 600 and we will say that, multiplying the root of 15

by 20 and the root of 40 makes the root of 6000 and the root of 600 and it is done.

6. Now I want to say it this way: I want to multiply 15 by 20 less the root of 40. So do it similarly to what you did before, that is, to multiply 15 by 20 which makes 300 and then you must multiply 15 by the root of 40. Bring 15 to the root making 225. Now multiply the added root of 225 by the subtracted root of 40 making the subtracted root of 9000 and you will say that multiplying 15 by 20 less the root of 40 makes 300 less the root of 9000 and it is done.

7. Now I want to say it this way: I want to multiply 5 and the root of 20 by 5 and the root of 20. Do it this way: put it in order as you see drawn below and then multiply 5 by 5, making 25 and save it. Now multiply by the cross which is 5 by the root of 20. You must bring 5 to the root, which is 25. Now multiply the root of 25 by the root of 20, which makes the root of 500.

5 plus $\mathfrak{R}$ of 20
via
5 plus $\mathfrak{R}$ of 20

Now multiply the other piece of the cross[1], that is also 5 by the root of 20, which also makes the root of 500. Now you must join the root of 500 with the root of 500. Always keep this in mind that when working to join two equal roots, always multiply the one part by four and that much is what the joining of those two roots make, which is the same as saying that multiplying 500 by 4 makes 2000 and you now have 25 and the root of 2000. Now it remains to multiply the root of 20 by the root of 20, which makes 20, which joined with 25 and the root of 2000, makes 45 and the root of 2000 and we will say that, multiplying 5 and the root of 20, [by 5 and the root of 20] makes 45 and the root of 2000 and it is done.

[1]The diagram in our scribe's source for rules 7 and 8 seems to have had a cross. See rule 13. He "cross-multiplies" the terms on the diagonals.

8. Now I want to say: I want to multiply 5 and the root of 20 by 5 less the root of 20.

5 plus 𝔑 of 20
via
5 less 𝔑 of 20

Again similarly, you place it as you placed the one before and then multiply also as you did the one before, that is, multiply 5 by 5 to make 25, and then multiply by the cross: 5 by the subtracted root of 20. You must bring 5 to the root, which is 25. Now multiply the root of 25 by the subtracted root of 20. It makes the subtracted root of 500. Now multiply the other the other piece of the cross, that is 5 by the added the root of 20 which is the added root of 500. Join it with 25 and the subtracted root of 500. They make exactly 25. Now it remains for you to multiply the added root of 20 and the subtracted root of 20 which is a subtracted 20. Between it and 25, there remains 5 and we will say that, multiplying 5 and the root of 20 by 5 [less] the root of 20 which makes 5 exactly and it is done.

9. Now I want to teach you how to divide the root of a number by a number and I want to say it this way: I want to divide the root of 300 by 4. I want to know how much it comes to. You must bring 4 to the root, that is, multiply 4 by itself. Divide 300 by 16 and it comes to the root of 18 and $\frac{3}{4}$ and we will say that dividing the root of 300 by 4 comes to the root of 18 and $\frac{3}{4}$ and it is done.

10. Now I want to say it this way: I want to divide 40 by the root of 30. I want to know how much it comes to. Again, do it similarly to what you did before, that is, bring 40 to the root which makes 1600. Now divide the root of 1600 by the root of 30. In truth it comes out as the root of $53\frac{1}{3}$ and you will say that, by dividing 40 by the root of 30 it comes out to the root of $53\frac{1}{3}$ and it is done.

11. Now I want to say it this way: I want to divide 20 and the root of 200 by 5. I want to know what it comes to. Do it this way: first divide the number which is 20 by 5. It comes to 4 and keep

it in mind. Now it remains for you to divide the root of 200 by 5, which makes you bring 5 to the root which is 25. Now divide 200 by 25. It comes to 8 and we will say that by dividing 20 and the root of 200 by 5, it comes to 4 and the root of 8 and it is done. And do the similar ones this way.

12. Now I want to say it this way: I want to divide 20 less the root of 300 by 5. I want to know how much it comes to. Again divide the number first, that is, 20 by 5. It comes to 4 and save it. Now it remains for you to divide the subtracted root of 300 by 5. Bring 5 to the root. It makes 25. Now divide 300 by 25. It comes to 12 and we say that dividing 20 less the root of 300 by 5 comes to 4 less the root of 12 and it is done.

13. Now I want to show you how to divide by two roots[2], which is to say, dividing a number by the root of number and a number. I want to say it like this: divide 100 by 10 and root of 20 and know what it comes to.

10 plus ℜ of 20

10 less ℜ of 20

These are the two terms, which are 10 and the root of 20, that you must bring into one and bring them in this way, which is, you will multiply 10 and the root of 20 against its residue–that where it says "plus" you put "less"– which is 10 plus the root 20 by 10 less the root 20 and, [with] this "less" that you put, it is the residue.

And now multiply 10 by 10 making 100, and then multiply via the cross, that is, 10 by the subtracted root of 20. You must bring 10 to the root which makes it 100. Now multiply the root of 100 by the subtracted root of 20 which is the subtracted root of 2000. Now multiply the other piece of the cross, that is, 10 by the added root of 2000. Join

[2]We think scribe erred and that the word should be "terms."

it to 100 less the root of 2000. It makes exactly one hundred. Now multiply the added root of 20 by the subtracted root of 20 but, since added by subtracted is subtracted, taken out of 100, there remains 80. Therefore we have that our divisor is 80, which we have gotten from the number and the root. Now you must always multiply that into which you want to divide by its residue, which, in this problem, is that you want to multiply 100 by 10 and the subtracted root of 20. Multiply 100 by 10, making 1,000 and keep in mind. Now multiply 100 by the subtracted root of 20; you must bring 100 into the root which makes it 10,000. Now multiply the root of 10,000 by the subtracted root of 20. They make the subtracted root of 200,000 and this, joined to 1000, makes 1000 less the root of 200,000 and this is what you want to be divided into 80 parts. First the number which is 1,000. It comes to 12 and $\frac{1}{2}$. Now it remains to divide the subtracted root of 200,000 by 80. You must bring 80 into the root, making 6,400. Now divide the subtracted root of 200,000 by the root of 6,400. In truth, it comes to the subtracted root of 31 and $\frac{1}{4}$. So we will say that, dividing 100 by 10 plus the root of 20, we get $12\frac{1}{2}$ less the root of $31\frac{1}{4}$. And so keep in mind to divide others, it is always to be considered which is more, the number or the root, so that the divisor has the larger one placed in front. Now consider which is more: either 10 or the root of 20, in which you see that the root of 20 is less than 5, therefore you always want to say the 10 first when it is more than the root.

14. And it is said to you: I want to divide 100 by 10 less the root of 30. I want to know how much it comes to.

10 less $\mathfrak{R}$ of 30

10 plus $\mathfrak{R}$ of 30

Again, in a similar way, you put the residue against the binominal, that is, you will multiply 10 less the root of 30 by 10 plus the root of 30, which is where you say less in the residue if it is plus and where you say plus in the residue if it subtracted from the thing that says plus. Therefore you must multiply 10 less the root of 30 by 10 plus the root of 30. And multiply via the cross the way you did before. That is, multiply 10 by 10 which makes 100. Now multiply 10 by the subtracted root of 30. You must bring 10 to the root which makes 100. Now multiply the root of 100 by the root of the subtracted root of 30 which makes the subtracted root of 3000. Now multiply the other piece of the cross, which is 10 by the added root of 30 which makes the added root of 3000. Joined to 100 and the subtracted root of 30, it makes exactly 100. Now multiply the subtracted root of 30 by the added root of 30 which makes a subtracted 30. When taken from 100, there remains 70 and 70 is our divisor so that two terms have been brought into one and it makes 70. Now, as we said in the previous division, you want to multiply that by which we want to divide by the residue. Therefore, we want to multiply 100 by 10 and the added root of 30 which, in truth, makes 1000 and the added root of 300,000. Now divide one thousand and added root of 300,000 by 70. First divide 1000. It becomes $14\frac{2}{7}$. Now you must bring 70 to the root, which is 4900. Now you divide the added root of 300,000 by the added root of 4900 [to get] the added root of 61 and $\frac{11}{49}$ and we will say that dividing 100 by 10 and the root of 30 gives $14\frac{2}{7}$ and the root of $61\frac{11}{49}$. And you should divide other quantities like so.

15. Now I want to tell you this way: I want to divide 100 by root of 60 less 5. I want to know how much it comes to.

ℜ of 60 less 5

ℜ of 60 plus 5

Again similarly, you must multiply the root of 60 less 5 by its residue, which is the root of 60 plus 5. Now multiply as you did before, that is, the root of 60 by the root of 60, which is 60; now multiply the root of 60 by the added 5; you must bring 5 to the root, which makes 25. Now multiply the root of 60 by the root of 25, making the added root of 1500, joined to 60, you have 60 and the added root of 1500. Now multiply the other piece of the cross, that is the root of 60 by the subtracted 5, which makes the subtracted root of 1500. Joined to 60 and the added root of 1500, they make 60. Now multiply the added 5 by the subtracted 5. They make a subtracted 25 that, taken from 60, leaves 35 and 35 is our divisor. Now you must multiply 100 by the residue of the root of 60 less 5, which is the root of 60 plus 5; you must to bring 100 to the root, making 10,000. Now multiply the root of 10,000 by the root of 60. They make the root of 600,000. Now multiply 100 by the added 5. They make the added 500. Now divide the root of 600,000 by 35. Bring 35 to the root, making 1225. Now divide the root of 600,000 by the root of 1225. It comes to the root of 489 and $\frac{39}{49}$. Now divide 500 by 35, it comes to $14\frac{2}{7}$. And so we have that, dividing 100 by the root of 60 less 5, it comes to the root of 489 and $\frac{39}{49}$ and plus the number $14\frac{2}{7}$ and it is divided. And you would follow this same method if one said the root 60 plus 5, as we said before.

16. Now I want to say it this way: I want to divide 40 and the root of 40 by 10 plus the root of 20. I want to know how much it comes to. First it is necessary to find the divisor in the way I said before which is that you multiply 10 plus the root of 20 against its residue, that is, against 10 less the root of 20, that in truth, you will find that the divisor will be 80 as we found it before. Now you must multiply 40 plus the root of 40 against the residue which is 10 less the root of 20 and keep with the previous method. You multiply 40 by 10 and less the root of 20, which is 4000 [Note: It should be 400] less the root of 32,000. You divide this by 80. You get 5 and $\frac{1}{4}$ less the root of 5. So we have already started with 400 and you get $5\frac{1}{4}$ less the root of 5. So we are left with the root of 40. So similarly, you multiply it against the residue, that is, against 10 less the root of 20. You must bring 10 to the root, which is 100. Now multiply the root of 100 by the root of 40, making the root of 4000. Now multiply the root of 40 by the subtracted root of 20 making the subtracted root of 800. Then you must divide the root of 4000 less the root of 800 by 80. Bring eighty to the root, making 6400. Now divide the root of 4000 by the root of 6400, getting the root of $\frac{5}{8}$. Now divide the root of 800 by the subtracted root

of 6400, get the subtracted root of $\frac{1}{8}$. Thus we have divided the root of 40 by 10 less the root of 20 getting the root of $\frac{5}{8}$ and the subtracted root of $\frac{1}{8}$. And at first we had started with 40 and there were $5\frac{1}{4}$ less the root of 5. Join it with root of $\frac{5}{8}$ and the subtracted root of $\frac{1}{8}$ you will have $6\frac{1}{4}$ plus the root of $\frac{5}{8}$ and the subtracted root of 5 and the subtracted root of $\frac{1}{8}$. And you must divide the others this way.[3]

Rules

1. **When the things are equal to the number,** you want to divide the number by the things and what it comes to is a number and that is the value of the thing.

I give you an example of the said rule and I want to say it this way: Make for me of 10 two parts such that dividing the greater by the lesser comes to 100. You must do it this way, that one part is a thing, which is the greater, therefore the other, which is the lesser, is the remaining up to 10, which comes to 10 less a thing. And this way we have supposed that the one that is the greater is a thing and the lesser is 10 less a thing. Next we want to divide a thing, which is the greater, by 10 less a thing, which is the lesser, and it must come to 100 and therefore it is necessary to multiply 100 by 10 less a thing, making 1000 less 100 things that are equal to one thing. Restore each part, which is to join 100 things to each part. You will have that 101 things are equal to a thousand numbers. Then we want to divide the number by the things which is 1000 by 101. It comes to 9 and $\frac{91}{101}$. And we supposed that the one part, which is the greater, was one thing, therefore the greater is 9 and $\frac{91}{101}$, and the lesser is the remainder up to 10, which is $\frac{10}{101}$ And so we have that the greater part of 10 is 9 and $\frac{91}{101}$ and the lesser is $\frac{10}{101}$.

Now I give you another example and I say it this way: There are three partners that have made a profit of 30 lire. The first put in 10 lire, the second put in 20 lire, and the third put that much into the company so that he took 15 lire from that profit. I ask

[3]The scribe made or copied a few errors. We leave it to the reader to find them.

how much the third put into the company and how much went to each of the other partners. If we want to know how much the third put into the company, one does it this way: suppose that the third had put one thing into the company. Next, you want to gather that which the first put in and that which the second put in and that which the third put in, that is, 10 and 20 and one thing. That makes 30 and one thing. And these 30 lire and one thing is what made a profit of 30 lire. Therefore, if you want to know what comes from this profit to the third, who we supposed put in a thing, know that you must multiply a thing by that which they profited and then divide among the entire company.[4] Therefore, we must multiply 30 by a thing, making 30 things, which is what must be divided by the entire company, that is, by 30 and a thing, from which comes what the third got. And we know he got 15 lire.[5] Therefore, you must divide 30 by 30 and a thing and it must come to 15 lire. So multiply 15 by 30 and a thing, making 450 and 15 things. Thus 450 numbers and 15 things are equal to 30 things. Restore each part, that is, take 15 things from each part. You will have that 15 things are equal to 450 numbers. Divide the number by the things, which is 450 by 15. It comes to 30 and this way we have that the value of the thing is 30. And we supposed that the third partner put one thing into the company. Therefore he put in 30. And if you want to know what went to the first and what to the second, remove 15, which is the profit of the third, from what profits they made, that is, from 30. There remains 15 and say: there are two partners that made a profit of 15. The first put in 10 and the second put in 20. What will be given to each one in truth is that to the first goes 5 to the second goes 10 and it is done.

2. **When the censi are equal to a number,** you want to divide the number by the censi and that which comes from it, its root, is the value of the thing.

Example: Find for me two numbers that are in proportion like

[4]Divide by the amount put in by the entire company.

[5]Note that $\frac{x}{x+30}$ is the percentage that the 3rd partner gets, so $\frac{30x}{x+30}$ is the amount he gets.

2 to 3 and, multiplied each by itself and taking one multiple from the other, be left with 20. I ask how much these numbers are. Suppose one number to be 2 things and the other to be 3 things which are indeed in proportion as 2 to 3. Next you want to multiply each by itself and take one multiple from the other to obtain 20. Therefore multiply the first number by itself which is 2 things by 2 things making 4 censi, and multiply the other number by itself which is 3 things by 3 things making 9 censi. Therefore you want to take the square of the first, which is 4 censi from the square of the second, which is 9 censi. There remains 5 censi. Therefore 5 censi are equal to the number 20, so you want to divide the number by the censi which is 20 by 5 that comes to 4. And it is what you take the root of, which is 2, and that is the value of the thing. And we supposed that one number was 2 things and we found that the value of the thing was 2. Therefore, the value of two things is 4 and that is the first number. And we supposed the other to be 3 things and the thing is worth 2 and the 3 things are worth 6. And so we have that one number is 4 and the other is 6 and it is done.

3. **When the censi are equal to the things,** you want to divide the things by the censi and that which it comes to is a number and the thing is worth that much.

Find me 2 numbers that are in proportion as 4 to 9 and, multiplying one against the other, make as much as when they are joined together. I ask how much is each number. Suppose that one number is 4 things and the other is 9 things, which are indeed in proportion as 4 is to 9. Therefore one number is 4 things and the other 9 things. So if we were to multiply 4 things by 9 things, it would make as much as joining 4 things and 9 things. Therefore multiplying 4 things by 9 things, makes 36 [censi], which are equal to 4 things joined with 9 things, which is to 13 things and therefore divide 13 by 36, which comes to $\frac{13}{36}$ things. We have that the thing is $\frac{13}{36}$. And we supposed that the one number was 4 things, therefore the number 4 by $\frac{13}{36}$ is 1 and $\frac{4}{9}$ and that much is the one number. And we supposed that the other was 9 things, therefore the other [is] 9 by $\frac{13}{36}$, which is 3 and $\frac{1}{4}$. And so we have that the one number is 1 and $\frac{4}{9}$ and the other is 3 and $\frac{1}{4}$. And we have quickly taught the three simple rules and we begin the other composite ones.

4. **When the censi and the things are equal to the number**, you want to divide by the censi, and then halve the things and multiply it by itself, and then join it onto the number. And the root of the sum less the half the things is the value of the thing.

   Someone lends another 100 lire for a term of 2 years, compounded at the end of the year, and when it comes to the end of two years 150 lire are returned to him. I ask at what rate per month was the lira lent. Suppose that the lira was lent at one thing of denari[6] per month. Therefore a lira makes 12 things of denari per year which is $\frac{1}{20}$ of a thing of lira. Therefore a lira makes $\frac{1}{20}$ things of lira per year so that 100 lire makes 100 by $\frac{1}{20}$ things of lire. And 5 things joined onto 100 lire makes 100 lire and 5 things and you have completed one year. And for the other year, multiply what comes at the end of the first year by the yield of the lira per year, that is, 100 plus 5 things by $\frac{1}{20}$ things which makes 5 things and $\frac{1}{4}$ censo which are what you want to join with 100 lire and 5 things, making 100 lire and 10 things and $\frac{1}{4}$ censo. And it is that much, between the interest and the capital, that 100 lire are [worth] in two years, which is 100 and 10 things and $\frac{1}{4}$ censo that are equal to 150 lire. Restore, that is remove 100 from each part, to get 10 things and $\frac{1}{4}$ censo equal to 50.

   Do it the way the rule says, that is to bring it to one censo, which is to divide by $\frac{1}{4}$, and you will have that one censo and 40 things are equal to 200 numbers. Halve the thing, it comes to 20. Multiply it by itself to make 400, join it to the number to make 600. Take the root which is a surd that presents as not a root.[7]

   Therefore we will say that the thing is worth the root of 600 less 20, which is the size of the thing. And we supposed that the lira was lent for a thing of denari per month. Therefore the lira was lent at the root of 600 denari less 20 denari per month and

[6]1 lira = 240 denari

[7]This means that 600 is not a perfect square.

it is done.

There are two men that have denari. The first says to the second, if you give me 14 of your denari that I will join with mine, I would have four times as much as you. The second one says to the first, if you give me the root of your denari, I would have 30 denari. I ask how much each had. First, let us suppose that the first had a censo and he asked for 14 from the second. Therefore he had a censo and 14 and must have four times as much as him. So it must be that what was left for the second is $\frac{1}{4}$ of a censo and $\frac{14}{4}$ which is $\frac{1}{4}$ censo and 3 and $\frac{1}{2}$. Therefore, before the second gave the first 14, he had $\frac{1}{4}$ censo and 17 and $\frac{1}{2}$. This way we have that the first man had a censo and the second one had $\frac{1}{4}$ censo and $17\frac{1}{2}$ numbers. Then the second one asked of the first for the root of his, which is the root of a censo which is a thing, to which you want to join to $\frac{1}{4}$ censo and 17 and $\frac{1}{2}$ numbers. This makes $\frac{1}{4}$ censo and a thing and 17 and $\frac{1}{2}$ numbers and with this you must have 30. So we have that $\frac{1}{4}$ censo and one thing and 17 and $\frac{1}{2}$ numbers are equal to 30. Restore each part which is to remove 17 and $\frac{1}{2}$ from each part and we get that $\frac{1}{4}$ censo and one thing is equal to 12 and $\frac{1}{2}$. Bring it to one censo to get that one censo and 4 things are equal to 50. Now halve the thing, it comes to 2. Now multiply it by itself making 4. Join it to 50 to make 54. We get that the value of the thing is the root of 54 less 2. [Multiply] by the root of 54 less 2, and the censo is worth that much which is in truth, 58 less the root of 864. And we supposed that the first had a censo. Therefore the first had 58 less the root of 864. Now know the second, which you supposed to have $\frac{1}{4}$ of a censo and 17 and $\frac{1}{2}$. So take $\frac{1}{4}$ of 58 less the root of 864 which is 14 and $\frac{1}{2}$ and the root of 54 onto which you had joined 17 and $\frac{1}{2}$ before to make 32 less the root of 54. And this way we have that the first has 58 less the root of 864 and the second man has 32 less the root of 54.

5. **When the things are equal to censi and a number,** you want to divide by the censi then halve the things and multiply it by itself and subtract the number. The root of that which remains plus the half the things is the value of the thing or, in truth, the half the things less the root of what remains.

Make for me of ten two parts such that, multiplying the larger against the smaller, makes 20. I ask how much each part is. Suppose that the smaller is a thing. Therefore the larger is the remaining up to 10 which comes to 10 less a thing. Next you must multiply the smaller, which is a thing, by the larger, which is 10 less a thing, making 10 things less a censo, the multiplication of which is equal to 20. Restore each part which is to join a censo to each part. You will have that 10 things are equal to a censo and 20 numbers. Bring to one censo and then halve the things, it comes to 5, and multiply it by itself, making 25. Subtract the number which is 20. There remains 5 of which you take the root, which clearly does not have a root. Therefore the value of the thing is that which was halved less the root of 5. And we had supposed that one part was a thing, the smaller one, thus the smaller is 5 less the root of 5 and the larger is what remains of 10 which comes to 5 plus the root of 5.

Someone makes two voyages.[8] The first makes a profit of 12. The second voyage was at that same rate. And when the ventures were completed they were found to be 54. We want to find out how much was moved [invested].

Let us suppose that he moves with a thing. In the first voyage he made a profit of 12. So at the completion of the first voyage he finds himself with 1 thing and 12. So it is clear that each thing makes him 1 thing and 12 in the first voyage. Therefore, if each one thing makes one thing and 12, how much each thing will make is 12. You must multiply one thing and 12 by one thing and 12 and divide by a thing.[9] One thing and 12 by one thing and 12 make a censo and 24 things and 144 numbers which you want to divide by a thing and this must come to 54. Therefore, multiply 54 by a thing, making 54 things, which is equal to a censo and 24 things and 144 numbers. Restore each part, which is to remove 24 things from each part. You will have that 30 things are equal to a censo and 144 numbers. Bring it to one censo, then halve the things, which is 15, and multiply it by

[8]The Italian word here is "viagi" or voyages. The voyages are two business ventures.

[9]In our notation, this is $\frac{(x+12)^2}{x}$. The rate is $\frac{12}{x}$ and $(x+12)+\frac{12(x+12)}{x} = \frac{(x+12)^2}{x}$.

itself which is 225. Remove 144, remaining with 81 of which you want to take its root and subtract it from 15. And so the value of the thing is 15 less the root of 81, which is 6. Thus 6 was moved and do similar ones this way.

And one could also say that one moved with the root of what remained plus the half of the thing, which is with the root of 81 plus 15, which is with 24. I said that to you, you would be as good in one way as in the other. And here's the proof: we say first one moves 15 less the root of 81, which is 6. The first investment yielded 12, therefore, upon completion of the first venture, he had 18. Therefore each 6 florins makes 18 for the first investment. Then 18 will be moved at the same rate. Multiply 18 by 18 and divide by 6, and it comes to 54, as was proposed. And if it was moved with the root of 81 plus 15, which is 24, it would also be good because the first investment made 12, and upon completion, you would have 36. Therefore each 24 makes 36. That which comes from 36 is invested at the same rate that you can clearly see makes 54 and this way you are good in one way as in the other.

Now, made this way, this rule is much to be praised because it gives two answers and you are as good with one as the other. But we must keep in mind that not all problems that use this rule can be answered with two answers if one does not check them.

And these are such that you must pick one answer or the other, which is to say that such problems make you answer that the value of the thing is the half of the things less the root of the remaining part or that you must say the root of the remaining part plus half the things. So, whenever such information comes to you, try one answer first, and if that does not come out to be true, certainly pick the other without doubting that you have the true answer. And keep in mind that this rule will require much discussion, but it won't stress me much because I have written much and it is enough.

I give you an example of what I said before and I say it this way. Make for me of 10 two parts such that, multiplied one against the other, and joined onto this multiplication the difference of one part from the other, makes 22. I want to know how much each part is. Suppose that one part is a thing and thus the other is 10 less a thing. Now multiply one against other, which is one thing by 10 less one thing, making 10 things less a censo. Now, after this multiplication, you want to join the difference which is one thing from 10 less a thing, making 10 less 2 things. To this difference, you want to join 10 things less a censo making 10 numbers and 8 things less a censo which is equal to 22. Restore each part. You will have eight things that are equal to 12 numbers and a censo. Divide by the censo and then halve the things, getting 4, multiplied by itself, making 16. Subtract the number which is 12 leaving 4. Take the root added to half the thing; it is the value of the thing. That is to say that the value of the thing is the root of 4 plus 4. And we had supposed that one part was a thing, therefore one part is the root of 4 plus 4 which is 6. Therefore, the other part is 4. Now try it and know if it is good. Multiply one against the other to make 24. Add the difference which gives 4 from 6 which is 2, making 26 which was supposed to make 22. Therefore we can see clearly that the value of the thing is not the radical of what remains plus half of the thing. So this was tried and not found true. For sure, choose the other answer and you will have the truth, that is, half the thing less the root of what remains, which is 4 less the root of 4, which is 2. Therefore, one part is 2 and the other is what remains from 10 which is 8. And this way we have that the smaller is 2 and the larger is 8.

6. **When the censi are equal to things and to the number**, you must divide by the censi and then take half the things and multiply by itself and join to the number. And the thing is worth the root of the sum plus the half the things.

   Someone has 40 gold florins and exchanges them for viniziani[10] and then from those viniziani he takes 60 and re-exchanges them for gold florins at one more viniziano per florin then had exchanged at first. And when he exchanged this way he found that, between the viniziani that remained when he had taken the 60

[10]Florins (fiori d'oro) and viniziani, usually spelled venetiani, are coins from Florence and Venice.

and the florins that he obtained from the 60 viniziani, joined together, made 100. I ask what the value of a florin is in viniziani. We suppose that the florin is worth a thing of viniziani. Therefore 40 gold florins are worth 40 things of viniziani. Next take 60 of these viniziani and re-exchange them for gold florins at a rate of one more viniziano per florin. Therefore, taking 60 viniziani from 40 things of viniziani, there remains 40 things less 60 viniziani and these are the viniziani that remain. Rejoined with what you got from the 60 viniziani, they make 100. Therefore, if we take 40 things less 60 viniziani from 100, you are left with what the 60 viniziani yielded in florins. So take 40 things less 60 from 100, and remain with 160 less 40 things. The florins that came from the 60 viniziani were 160 less 40 things and these were exchanged at a rate of one more viniziano per florin than before. So changing the 60 viniziani at a rate of one thing and a viniziano, we have that 60 viniziani are worth, in gold florins, 160 less 40 things. So we need to know that 160 less 40 things of florins, at a rate of one thing and one viniziano, are worth 60 viniziani. Therefore, multiply 160 less 40 things by one thing plus one, making 120 things less 40 censi plus 160 numbers that are equal to 60 viniziani. And this way, we have that 120 things less 40 censi plus 160 are equal to 60. Restoring each part, you will have that 40 censi are equal to 120 things and 100 numbers. Divide by the censi. You have that the censo is equal to 3 things and 2 and $\frac{1}{2}$ numbers. Halve the thing, to get $\frac{3}{2}$, multiply it by itself, making 2 and $\frac{1}{4}$. Join to the number to get 4 and $\frac{3}{4}$. We have that the value of the thing is the root of 4 and $\frac{3}{4}$ plus half the thing which was $\frac{3}{2}$. And we supposed that the florin is worth one thing of viniziani, therefore the florin is worth the root of 4 and $\frac{3}{4}$ [plus 1 and $\frac{1}{2}$].

7. **When cubes are equal to a number**, you want to divide the number by the cubes and the cube root of that which it comes to is the value of the thing.

   Find 3 numbers that are in proportion as 2 to 3 and as 3 to 4 and, multiplying the first by the second and then against the third, the total makes 96. Suppose that one is 2 things, and the other is three things and the other is 4 things. Multiply 2 things by 3 things. They make 6 censi. Now multiply 4 things by 6 censi. They make 24 cubes. And we want 96. So we have that 24 cubes are equal to 96 numbers. Divide by the cubes to get 4. The cube root of 4 is the value of the thing. And you supposed that one was 2 things. Bring 2 things to the cube root, making 8, multiplying 8 by 4, that makes 32 and one is the cube root of 32. The other you supposed to be 3 things. Bring it to the cube root making 27, multiplying 27 by 4 makes 108. The second is the cube root of 108. The third you supposed to be 4 things. Bring 4 to the cube root which is the cube root of 64, multiply by 4 making 256 and the third is the cube root of 256 and you are done.

8. **When the cubes are equal to things**, you want to divide the things by the cubes and the root of what it comes to is the value of the thing. Find for me two numbers such the parts are to one another as 2 to 3. And multiplying the first by itself and then by the number, make as much as the other number. Suppose that one is 2 things and the other 3. Multiply 2 things by 2 things, making 4 censi. Multiply 4 censi by 2 things, making 8 cubes. We have that 8 cubes are equal to 3 things. You must divide the things by the cubes, that is, 3 by 8 which comes to $\frac{3}{8}$. We have that the thing is worth the root of $\frac{3}{8}$. And you supposed that one was two things. You must bring 2 things to the root and say 2 by 2 make 4. Multiply the root of 4 by the root of $\frac{3}{8}$ to get the root of one and $\frac{1}{2}$. We have that one number is the root of one and $\frac{1}{2}$. And you supposed that the other is 3 things, brought to the root makes 9. Multiply the root of 9 by the root of $\frac{3}{8}$ to get the root of 3 and $\frac{3}{8}$. We have that the first number is the root of one and $\frac{1}{2}$ and the second the root of 3 and $\frac{3}{8}$.

9. **When the cubes are equal to censi**, you want to divide the censi by the cubes and that which results is a number and that is the value of the thing. Find for me two numbers that are in proportion one part to the other as 3 is to 4, and multiplying the first by itself and then by the number, and the other by itself. Suppose that one is 3 things and the other is 4 things. Multiplying 3 things by 3 things make 9 censi and 3 things by 9 censi make 27 cubes. Now multiply the other number by itself, that is, 4 things by 4 things, making 16 censi. We have that 27

cubes are equal to 16 censi. You must divide the censi by the cubes, that is 16 by 27, resulting in $\frac{16}{27}$. We have that the value of the thing is $\frac{16}{27}$. And you supposed one to be 3 things. Multiply 3 things by $\frac{16}{27}$, making 1 and $\frac{7}{9}$. We have that one number is 1 and $\frac{7}{9}$. The other you supposed to be 4 things. Multiply 4 by $\frac{16}{27}$, making 2 and $\frac{10}{27}$ and that is the second number and it is done.

10. **When the cubes are equal to censi and to things,** you must divide by the cubes and then halve the censo and multiply it by itself and join it to the thing and the root of the sum plus half the censo is the value of the thing.

    Find for me three numbers that are in proportion together as 3 to 4 and 4 to 5 and, multiplying the first by itself and then by the number, makes the same as the second multiplied by itself and then put it onto the third number. We ask, what are the numbers. Suppose that one is 3 things, the other is 4 things, and the other is 5 things. Now three things by three things make 9 censi and three things by 9 censi make 27 cubes. Multiplying 4 things by 4 things makes 16 censi. Put onto it 5 things and you have 16 censi and 5 things. We have that 27 cubes are equal to 16 censi and 5 things. You must divide by the cubes, resulting in one cube equal to $\frac{16}{27}$ censi and $\frac{5}{27}$ things. Halve the censo and multiply it by itself making $\frac{64}{729}$. Put it onto the thing, making $\frac{199}{729}$. [The root of $\frac{199}{729}$] plus the half of the censi is the value of the thing. You must make 3 things and 4 things and 5 things that you had supposed were the numbers and do it like I taught you before.

11. **When the cubes are equal to the root of a number**, you must divide the root of the number by the cubed thing and that which results from it is the root of the cube root, which is the value of the thing.

    Find for me 3 numbers that are in proportion together as 2 to 3 and 3 to 4 and, multiplying the first by the second and then by the third, makes the root of 12. Suppose that the first is 2 things, the other is 3 things, and the other is 4 things. Multiply 2 things by 3 things and then by 4 things making 24 cubes. They are equal to the root of 12. Bring 24 to the root making

576. Divide 12 by 576 resulting in $\frac{1}{48}$ and the root of the [cube] root of $\frac{1}{48}$ is the value of the thing. And you supposed one to be 2 things. Bring 2 things to the root of the cube root making 64. Multiplying 64 by $\frac{1}{48}$ makes the root of the cube root of 1 and $\frac{1}{3}$ and that is the number. The other you supposed to be 3 things. Bring 3 to the root of the cube root making 729. Multiplying 729 by $\frac{1}{48}$ makes 15 and $\frac{3}{16}$. We have that the second number is the root of the cube root of 15 and $\frac{3}{16}$. And you supposed the other to be 4 things. Take 4 to the root of the cube root making 4096. Multiply 4096 by $\frac{1}{48}$ to make 85 and $\frac{1}{3}$ and the third is the root of the cube root of 85 and $\frac{1}{3}$ and it is done.

12. **When the cubes are equal to things and a number,**[11] you must divide by the cubed things, then halve the things and multiply it by itself and put it with the number and the root of that plus half of the things is the value of the thing. Find for me two numbers such that the parts, one to another, are as 2 to 3 and multiplying the first by itself and then by that number, makes as much as the other plus 12. Do it this way and suppose that one is 2 things and the other is three things. Multiplying 2 things by 2 things makes 4 censi. Multiplying 2 things by 4 censi makes 8 cubed things and these must be 3 things and 12. We have that 8 cubed things are equal to three things and 12 numbers. You must divide by the cubes and get that one cube is equal to $\frac{3}{8}$ of the thing and 1 and $\frac{1}{2}$ number. You must halve the thing, which is $\frac{3}{16}$. Multiplying it by itself makes $\frac{9}{256}$. Put it with the number, it makes 1 and $\frac{274}{512}$. We have that the value of the thing is the root of 1 and $\frac{274}{512}$ plus $\frac{3}{16}$ for the number. You must make 2 things and then 3 things and each is that much and it is done.

13. **When the cubes are equal to censi and a number,**[12] you must divide by the cubes and then halve the censi and multiply it by itself and put it onto the number and the root of that which is made plus the half of the censi is the value of the thing.

Find for me 2 numbers such that the parts are to one another

[11]This example is incorrect! The cubic equation is treated as a quadratic.

[12]This example is also incorrect. Again the cubic equation is treated as a quadratic.

as 2 to 3 and multiplying the first by itself and then by the number, make as much as the second multiplied by itself plus 12. Do it this way. Suppose that one is 2 things and the other 3 things. Multiplying 2 things by 2 things makes 4 censi and multiplying 2 things by 4 censi makes 8 cubes. Now say 3 things by 3 things make 9 censi, put onto 12 and you have 9 censi and 12 numbers. We have that 8 cubes are equal to 9 censi and 12 numbers. Now divide by the cubed things, and verify that one cube is equal to 1 and $\frac{1}{8}$ of a censo and 1 and $\frac{1}{2}$ number. Halve the censi which comes to $\frac{9}{16}$, multiplied by itself, makes $\frac{81}{256}$, put it with the number that is 1 and $\frac{1}{2}$, makes 1 and $\frac{209}{256}$. And the root of 1 and $\frac{209}{256}$ plus the half of the censo is the value of the thing. And to make it so that it comes to 2 things and to three things, you must bring them to the root like you did before and such are the numbers.

14. **When the cubes and the things are equal to censi**, you must divide by the cubes and then halve the censi and multiply it by itself and subtract the things and that which remains, its root plus the half the censi, is the value of the thing. Or in truth, half the censi less the root of that which remains is the value of the thing. Find for me 3 numbers that are together in proportion as 3 is to 4 and 4 is to 5. Multiply the first by itself and then by the number and this multiplication put onto the second number makes as much as the third number multiplied by itself. Do it this way: suppose one number to be 3 things and the other 4 things and the other 5 things. Now multiply 3 things by 3 things, making 9 censi and multiply 9 censi by 3 things, making 27 cubes and put it onto the second thing that you supposed was 4 things. And you have 27 cubes and 4 things. Multiply 5 things by itself, making 25 censi. We have 27 cubes and 4 things are equal to 25 censi. You must divide by the cube to get that one cube and $\frac{4}{27}$ things are equal to $\frac{25}{27}$ censi. Half the censi, which is $\frac{25}{54}$,[13] subtract the thing which is $\frac{4}{27}$. There remains $\frac{17}{54}$. We have that the value of the thing is the root of $\frac{17}{54}$ plus the half the censo which is $\frac{25}{54}$. And you supposed one to be 3 things, bring it the root, so multiply 3 by 3 to make 9, multiply 9 by $\frac{17}{54}$ making 2 and $\frac{5}{6}$. And 3 things by $\frac{25}{54}$ make 1

[13]The author forgets to square $\frac{25}{54}$ and continues with this mistake.

thing and $\frac{7}{18}$. We have that the first number is the root of 2 and $\frac{5}{6}$ plus 1 and $\frac{7}{18}$. You must do 4 things for the second where the value of the thing is the root of $\frac{7}{54}$ plus the number $\frac{25}{54}$. And then you must do 5 things for the third. By the method given for the above, that is the way you will do similar problems.

15. **When the cubes and the censi are equal to the things**, you want to divide by the cubed things and halve the censi and multiply it by itself and join it to the things. And the root of the sum less the half of the censi is the value of the thing.

16. **When the censi are equal to the cubes and to the things**, you must divide by the cubes and then halve the censi and multiply it by itself, subtract the things. And the root of that which remains plus the half of the censi is the value of the thing, or in truth, the half of the censo less the root of the remaining.

17. **When the censi di censi are equal to the number**, you want to divide the number by the censi di censi and the root of the root is the value of the thing.

18. **When the censi di censi are equal to the things,** you want to divide the things by the censi di censi and the cube root of that is the value of the things.

19. **When the censi di censi are equal to censi**, you want to divide by the the censi di censi and the root of that is the value of the thing.

20. **When the censi di censi are equal to cubes**, you want to divide the cubes by the censi di censi and what comes is a number and that is the value of the thing.

21. **When the censi di censi and the cubes are equal to censi**, you want to divide by the censi di censi and then halve the cubes and multiply by itself and join it to the censi and the root of the sum less half the cubes is the value of the thing.

22. **When the cubes are equal to censi di censi and to censi**, you must divide by the censi di censi and then halve the cubes and multiply by itself and subtract the censi and the root of the

sum plus the half of the cube is the value of the thing, or in truth, the half of the cubed thing less the root of what remains.

23. **When the censi di censi are equal to cubes and to censi**, you must divide by the censi di censi and then take half the cubes and multiply by itself and join it with the censi and the root of the sum plus half the cubes is the value of the thing.

24. **When censi di censi and censi and are equal to the number**, you must divide by the censi di censi and then halve the censi [and multiply by itself] and join it to the number and subtract the half the censi. The root of what remains is the value of the thing, but actually, the root of the root of the sum less half the censo is the value of the thing.

Problems

1. Make for me of 10 two parts such that, multiplied one against the other, they make 22. I want to know how much each part is. Do it this way: Suppose that one part is a thing so it must be that the other is 10 less a thing. Now multiply one thing by 10 less one thing. They make 10 things less a censo and these must make 22. So 10 things less a censo is equal to 22. Now restore each part so that you put a censo to each part. You will have that 10 things are equal to 22 numbers and a censo. Divide by the censi, which is by one. It comes to 10 things [equal to] 22 numbers and a censo. Now you must halve the things which are 10, its half is 5, multiply it by itself, making 25, subtract the number which is 22. There remains 3 and say that the thing is worth half the things, which was 5, plus the root of 3. And we supposed that one part was a thing so one part is 5 plus the root of 3 and we supposed that the other part was the rest of 10, which comes to 5 less the root of 3. And this way we have that one part is 5 plus the root of 3 and the other is 5 less the root of 3 and it is done.

2. Make for me of 10 two parts such that, multiplying each by itself and joined together, these two multiplications make 60. I want to know how much each part is. Do it this way: suppose that one part is a thing and therefore it must be that the other is 10 less a thing. Now multiply each by itself and say: one thing by one thing makes a censo. Now multiply 10 less a thing by itself. Do 100 less 10 things. Now say: one subtracted thing by 10 and a subtracted thing makes 10 subtracted things plus a censo. You have 100 numbers and 20 subtracted things plus a censo, joined with the censo from above, and with 100 numbers less 20 things plus 2 censi, they are equal to 60 numbers. Restore each part, that is, you put 20 things on each part. You will have that 100 and two censi are equal to 60 plus 20 things. Now take 60 numbers from each part. You will have that 40 numbers and 2 censi are equal to 20 things. Now divide each part by the censi, that is by 2. You will have that 10 things are equal to 20 numbers and a censo. Now halve the things that are 10, the half of which is 5, and 5 multiplied by itself makes 25. Take

away the number which is 20 and 5 remains. And say the value of the thing is 5 plus the root of 5 that remained. We supposed that one part was a thing therefore one part is 5 plus the root of 5 and that the other is what remains of 10, which is 5 less the root of 5. Now I want to teach by an easier method and I say it like this: that you always halve the 10 and then multiply it by itself, making 25 and from this 25, take the double of 10 which is 20. Subtracted from 25, there remains 5 and I will say that one part is the half of 10, which is 5, plus the root of 5 and the other part is 5 less the root of 5.

3. Make for me of 10 two parts such that multiplied one against the other and joined onto the difference of one part from the other makes 24. I want to know how much each part is. Suppose that one part is a thing, therefore it must be that the other is the remainder up to 10 which becomes 10 less a thing. Then you must multiply a thing by 10 less a thing, which makes 10 things less a censo. Now join the difference that you have from a thing up to 10 less a thing. If you want to know what the difference comes to, subtract one thing from 10 less a thing. There remains 10 less 2 things and join these with 10 things less one censo, making 10 numbers and 8 things less a censo and this is equal to 24 numbers. Now restore each part, which is that you put a censo onto each part. You will have that 10 numbers and 8 things are equal to 24 and a censo. Now subtract 10 numbers from each part. You will have that 14 numbers and a censo are equal to 8 things. Therefore you must halve the thing, the half of which is 4. Multiply it by itself making 16, subtract 14 and 2 remains. And say that the value of the thing is 4 less the root of 2. And we had supposed that one part was the thing. Therefore one part is four less the root of 2 and it must be that the other is the remainder up to 10 which is 6 plus the root of 2. And this way we have that one part is 4 less the root of 2 and the other is 6 plus the root of 2 and it is done.

4. Make for me of 10 two parts such that, dividing the larger by the smaller, comes to the difference that one part has from the other. Suppose that the smaller part is a thing. The other part comes to be 10 less a thing which is the larger part. Then you

must divide 10 less a thing by a thing and it must come to the difference. Go to the other part which is a thing up to 10 less a thing. To know what the difference is, I give a material example and I say it this way. I want to know the difference that 12 has from 30. Do it this way: always take the smaller number from the larger and the remainder that remains is the difference that 12 has from 30. Therefore you must take 12 from 30, remaining with 18 and 18 is the difference from 12 to 30. It is in this same manner that you must do our problem. Therefore you must take one thing from 10 less a thing. There remains 10 numbers less two things. So you must divide 10 numbers less a thing by a thing and say it comes to 10 numbers less 2 things. Then multiply a thing by 10 less 2 things making 10 things less 2 censi and this is equal to 10 numbers less one thing. Restore each part, which is that you give two censi to each part. You will have that 10 things are equal to 10 numbers less one thing plus 2 censi. Again, restore each part of the things, which is that you must add one thing to each part. You will have that 11 things are equal to 10 numbers plus 2 censi. Thus we have brought this problem to the fifth rule which says that you must divide by the censi so that in this problem, you want to divide by 2. You will have that 5 numbers and one censi are equal to 5 and $\frac{1}{2}$ things. Now you must halve the things so that we do it the way the rule says. The things are 5 and $\frac{1}{2}$, the half of which is 2 and $\frac{3}{4}$. Multiply it by itself making 7 and $\frac{9}{16}$. From this you want to take the number which is 5. Remain with 2 and $\frac{9}{16}$ and say that the value of the thing is the root of two and $\frac{9}{16}$ plus half the things which were 2 and $\frac{3}{4}$ and we supposed that the smaller part was a thing and therefore the smaller part is 2 and $\frac{3}{4}$ plus the root of 2 and $\frac{9}{16}$. And the larger part is the remainder up to 10 which comes to 7 and $\frac{1}{4}$ less the root of 2 and $\frac{9}{16}$.

5. Make for me of 10 two parts such that dividing the larger by the difference that one part has to the other comes to the smaller part. I want to know how much each part is. Suppose the smaller part to be a thing. The larger becomes what remains up to 10 which becomes 10 less a thing. Now you must divide 10 less a thing by the difference that the smaller has from the larger. To know the difference that one thing has from 10 less a

thing, you always take the smaller part from the larger, which is one thing from 10 less a thing leaving 10 numbers and less two things. And now you must divide 10 numbers less a thing by 10 numbers and less 2 things and it must come to the smaller part which is a thing. Now I want to pose to you a very clear example and I say it this way: I want to divide 24 by 4 that I know really must come to 6. I say that if you multiply 6, which it has to come to, against the divisor, that is, against 4, that it will make the quantity that we want to divide which is 24. We return to our problem, which is that we must divide 10 numbers less one thing by 10 numbers less two things and it must come to a thing. So you must multiply one thing by 10 less two things, which makes 10 things less two censi. This is equal to that which we want to divide, that is, to 10 numbers less one thing and this way we have that 10 things less 2 censi are equal to 10 numbers less one thing. Now restore each part which is that you give 2 censi to each part. You will have that 10 things are equal to 10 numbers less one thing plus two censi. Again restore each part which is that you give to each part one thing. You will have that 11 things are equal to 10 numbers plus two censi. This is the [fifth] rule which says to divide each thing by the censi which are two. You will have that $5\frac{1}{2}$ things are equal to 5 numbers plus one censo. Now you must halve the things which are $5\frac{1}{2}$, the half of which is $2\frac{3}{4}$, multiply it by itself, making $7\frac{9}{16}$. From this you want to take away the numbers which are 5, remaining with $2\frac{9}{16}$. We have that the value of the thing is the half which was $2\frac{3}{4}$ plus the root of what remains which was 2 and $\frac{9}{16}$. And we supposed that the smaller was a thing, and so the smaller part is $2\frac{3}{4}$ plus the root of 2 and $\frac{9}{16}$ and the larger part is what remains of 10, which comes to 7 and $\frac{1}{4}$ less the root of two and $\frac{9}{16}$. And this way we have that the smaller part is $2\frac{3}{4}$ plus the root of 2 and $\frac{9}{16}$ and the larger is $7\frac{1}{4}$ less the root of 2 and $\frac{9}{16}$ and it is done. You must do similar problems this way.

6. Make for me of 10 two parts such that multiplying each part by itself and joining these two multiplications together and dividing by the difference of one part to the other comes to 25. I want to know how much each part is. Do it this way: suppose that one part is a thing. The other comes to 10 less a thing. Now

multiply each one by itself and say one thing by one thing which makes one censo. Now multiply the other and say 10 and less a thing by 10 and less a thing makes 100 numbers less 20 things plus one censo, joined with a censo, makes 100 numbers less 20 things plus two censi. And this you must divide by the difference that the smaller part has from the larger. So you take one thing from 10 less a thing. There remains 10 numbers and less two things. And these 10 numbers less two things are the difference. Therefore you must divide 100 numbers less 20 things plus two censi by 10 numbers less 2 things and it must come to 25. So multiply 25 against the divisor, that is, against 10 numbers and less two things. It makes 250 numbers less 50 things and this is equal to 100 numbers less 20 things plus 2 censi. Now you must restore each part, that is, you put 50 things on each part. You will have [100] numbers plus 30 things plus two censi equal to 250 numbers. Now take 100 numbers from each part. You will have that 30 things plus two censi are equal to 150 numbers. And this is the [fourth] rule which is that you must divide by the censi, that is by one.[14] By dividing by one, the thing that you want to divide comes to that much. Now you must halve the things which are 30, the half of which is 15, multiply it by itself making 225, join it to 150. You will have 375 and I say that the value of the thing is the root of 375 less the half of the thing which was 15. The other part is what remains from 10 which comes to 25 less the root of 375. And do similar ones this way.

7. Make for me of 10 two parts such that, multiplying one by 3, makes as much as the other multiplied by the root of 8. I want to know how much each part is. Suppose that one part is a thing. The other comes to be 10 less a thing. Now multiply one thing by 3, making 3 things. Now you must multiply the other by the root of 8. Therefore you must bring 10 less a thing to the root which makes 100 numbers less 20 things and plus a censo. Now multiply the root of 8 by the root of 100 numbers less 20 things plus a censo which makes 800 numbers less 160 things plus 8 censi and this is equal to 3 things which is the multiplication from the first part. Therefore you must bring 3

[14]This should be two. The mistake propagates so the answer is incorrect.

things to the root, which make 9 censi. Therefore we say that 9 censi are equal to 800 numbers less 160 things plus 8 censi. Restoring the parts, which is that you give 160 things to each part and take 8 censi from each part. You will have that 160 things plus one censo are equal to 800 numbers. Bring it to one censo, you will have that 160 things [plus one censo] are equal to 800 numbers. Halve the things that are 160, the half is 80. Multiply it by itself, making 6400. Join it to the number which is 800 making 7200 and say it this way: one part is the root of 7200 less 80. The other part is what remains up to 10 which comes to 90 less the root of 7200. Therefore the thing is worth the root of 7200 less 80. And we supposed that one part was a thing, so that one part is the root of 7200 and less 80. The other is 90 less the root of 7200.

8. Make for me of 10 two parts such that, multiplying the larger by itself, adding the difference that one part has to the other, and dividing by the smaller part, it comes to 40. I want to know how much each part is. Suppose that the larger part is a thing. The smaller comes to what remains up to 10, which is 10 less a thing. Now multiply a thing by itself, making a censo. Join the difference that is from a thing to 10 less a thing, that is, the smaller which is 10 less a thing from the larger which is a thing, remain with 2 things less 10. Joined with a censo, you have a censo and 2 things less 10 and this you want to divide by the smaller part, which is 10 less a thing, and it must come to 40. So if you multiply 40 by 10 less a thing, it makes a censo plus 2 things and less 10 numbers. Then 40 by 10 less a thing makes a censo plus 2 things less 10 numbers. Therefore 40 by 10 less a thing makes 400 numbers less 40 things and this is equal to a censo plus 2 things less 10 numbers. Restoring the parts, which is to give 10 numbers to each part and then 40 things to each part, you will have that a censo and 42 things are equal to 410 numbers. Divide by the censi, which is one, that comes to the same, halve the things, which are 21, multiply it by itself, making 441, join it to the number, making 851 and we say that the thing is worth the root of 851 less 21. And we suppose that the larger was a thing, therefore the larger part is the root of 851 less 21 and the smaller part is the remainder up to 10, which

is 31 less the root of 851. And this way we have that one part, which is the larger, is the root of 851 less 21 and the other, which is the smaller, is 31 less the root of 851 and it is done.

9. Make for me of 10 three parts in continued proportion[15] such that multiplying the first against the second makes 4 and multiplying the second against the third makes 8. I want to know how much each part is. You must consider this problem in this way and say that multiplying the first against the second makes 4 and multiplying the second against the third makes 8. I say that clearly one sees that the third must be twice as much as the first since multiplying the third against the second makes twice as much as multiplying the first against the second.

   And when the parts are in proportion, multiplying the first by the third makes as much as multiplying the second by itself. Now suppose the first part to be a thing and I showed you before that the third had to be twice as much as the first. Therefore the third must be two things. Now we have that the first part is a thing and the third part is two things, therefore we have that the second is the remainder up to 10 which is 10 less three things. Now multiply the first, which is a thing, against the second, which is against 10 less three things which makes 10 things less 3 censi and this must make 4. Therefore 10 things less 3 censi are equal to 4. Restore each part, so that you give 3 censi to each part. You will have that 10 things are equal to 3 censi and 4 numbers. You must divide by the censi, which are 3. In this problem, you will have that $3\frac{1}{3}$ things are equal to one censo and one and $\frac{1}{3}$ number. Now you must halve the things, the half of which is one and $\frac{2}{3}$, multiply it by itself, making 2 and $\frac{7}{9}$, subtract the number, which is 1 and $\frac{2}{3}$, remaining with one and $\frac{4}{9}$. And we say that the thing is worth the half, which was $1\frac{2}{3}$ and the root of 1 and $\frac{4}{9}$. And we supposed that the first part was a thing, therefore the first part is one and $\frac{2}{3}$ plus the root of one and $\frac{4}{9}$. And the second part was 10 less 3 things. So to know what three things are worth, you must multiply 3 by one and $\frac{2}{3}$, which makes 5. Now multiply 3 by the root of one and $\frac{4}{9}$, so that you must bring the 3 to the root, which is 3 by

[15]Three numbers x, y and z are in continued proportion if x is to y as y is to z.

3, making viiii [9]. Now multiply 9 by $1\frac{4}{9}$, making the root of 13. So do 5 less the root of 13 subtracted from 10, and indeed there remains 5 less the root of 13 and it is done.

10. Make for me of 10 three parts in continued proportion such that multiplying each part by itself and joining them together, these three multiplications make 40. I want to know how much each part is. What is considered in this problem is this, that the square of the first and of the third with the square of the second removed, leaves the squares of all three parts.[16] And so suppose that the second part is a thing. Therefore 10 less a thing remains between the first and the third. Now multiplying 10 less a thing by itself, as you learned from what was considered, makes 100 numbers less 20 things plus a censo. Now multiply a thing, which is the second part, making a censo, subtract from the square of the first and the third, there remains 100 numbers less 20 things and this is equal to the squares of all three things, which is 40.[17] Restore the parts which is that you give 20 things to each part. You have 20 things and 40 numbers equal to 100 numbers. Subtract 40 numbers from each part. You will have that 20 things are equal to 60 numbers. Divide 60 by the things, which is by 20, and it comes to 3. We have that the thing is worth 3. And we supposed that the second part was a thing therefore the second part is 3. Therefore 7 remains between the first and the third parts, which is what remains from 10. Now you need to find each of the first and third, and to find that, multiply 3 by itself, making 9, take it from 40 and there remains 31 which is what the parts must come to. Now you must make 7 into two parts that, multiplying each by itself and joining them together, these two multiples make 31. I want to know how much each part is. Suppose that the larger part is a thing so it must be that the smaller is 7 less a thing. Now multiply each by itself and join them together. Indeed you will find that they make 49 numbers less 14 things plus 2 censi. And all of this is equal to 31. Restore the parts, which is that you give 14 things to each part. You will have 14 things and 31 numbers are equal to 49 numbers plus two censi. Taking 31 numbers from each

---

[16] $(x+z)^2 - y^2 = x^2 + y^2 + z^2$ because $xz = y^2$.

[17] $(x+z)^2 = (10-y)^2$ and so $(10-y)^2 - y^2 = 40$.

part, you will have that 14 things are equal to 18 numbers and 2 censi. Divide by the censi, which are 2, and get that 7 things are equal to 9 numbers and 1 censo. Now you must halve the things, which are 7, the half is $3\frac{1}{2}$, multiply it by itself, making $12\frac{1}{4}$, take away the number, which is 9, remain with $3\frac{1}{4}$. And you say that the thing is worth the half, which was $3\frac{1}{2}$ plus the root of $3\frac{1}{4}$. And we have supposed that the larger was a thing. Therefore the larger part is $3\frac{1}{2}$ plus the root of $3\frac{1}{4}$ and the other part is the remainder up to 7 which is $3\frac{1}{2}$ less the root of $3\frac{1}{4}$. The second part is 3, the third part is $3\frac{1}{2}$ plus the root of $3\frac{1}{4}$ and it is done.

11. Make for me of 20 two parts such that taking the root of each part and joining together makes 5. I want to know how much each part is. Consider this problem this way: If the 2 roots of 20 must make 5, then if you make 5 into two parts such that multiplying each by itself and joining makes 20,[18] then as much as each part will be, that quantity will be the root of each part that makes 20. Now make 5 into two parts such that multiplying each by itself and joining together makes 20. Do it this way: halve 5, its half is $2\frac{1}{2}$ multiply it by itself, making $6\frac{1}{4}$, remove $2\frac{1}{2}$, remain with $3\frac{3}{4}$ and I will say that one part of 5 is $2\frac{1}{2}$ plus the root of $3\frac{3}{4}$ and the other is $2\frac{1}{2}$ less the root of $3\frac{3}{4}$. Therefore, according to the consideration given before, the root of one part of 20 will be $2\frac{1}{2}$ plus the root of $3\frac{3}{4}$ and the root of the other part of 20 will be $2\frac{1}{2}$ less the root of $3\frac{3}{4}$. To know each part of 20 one does it this way, which is, that you multiply $2\frac{1}{2}$ plus the root of $3\frac{3}{4}$ by itself, which makes 10 plus the root of $93\frac{3}{4}$. Now you must multiply the other part, which is $2\frac{1}{2}$ less the root of $3\frac{3}{4}$ by itself, which makes 10 less the root of $93\frac{3}{4}$ and this way we have that one part of 20 is 10 plus the root of $93\frac{3}{4}$ and the other is 10 less the root of $93\frac{3}{4}$. And this is how you must respond to this problem.

12. Divide 100 by a quantity and that which it comes to, multiplied by 8, makes the divisor. I want to know the divisor. I give

[18] $\sqrt{x} + \sqrt{20 - x} = 5$. Let $y = \sqrt{x}$ and $5 - y = \sqrt{20 - x}$. It works because $y^2 + (5 - y)^2 = x + (20 - x) = 20$. In the next part, the scribe skips steps but the results are correct.

you an example first and I say it this way: I want to divide 24 by 12 and that which it comes to, multiplied by 6, makes the divisor. I say that if you multiply 6 by 24 and divide by 12 that it comes to the divisor. Therefore let us suppose that we divide 100 by a thing. I say that if you multiply 8 by 100, divide by a thing, which is 800 divided by a thing, then you have that 800 divided by a thing must come to the thing. Now I give you an example at this point and I say it this way: If I divide 24 by 4, that I know comes to 6, I say that if I multiply it by the 6 that resulted by 4 which was the divisor, I say that it will make the 24 that I divided. Now I return you to our problem, which was that we had divided 800 by a thing and it must come to a thing. Therefore you multiply a thing by a thing, which makes a censo, and I say that a censo is equal to 800 numbers. And the rule says that when the censi are equal to a number that you want to divide the number by the censi and the root of that which results is the value of the thing. Therefore, divide 800 by one censo which comes to 800 and the root of 800 is the value of the thing. And we supposed that we divided 100 by a thing. Therefore we divide 100 by the root of 800.

13. Someone divided 100 by a quantity and then divided 100 by 5 more than the first and joining these two events together makes 20. I want to know by how much 100 is divided and by how much the next is divided. Suppose that you divide 100 by a thing. It comes to 100 divided by a thing. And then say that 100 is divided by 5 more than the first. So you must divide 100 by a thing and 5. It comes to 100 divided by a thing and 5. Now you need to join 100 divided by a thing with 100 divided by a thing and 5. Now I want to show you something similar that is good advice for this joining and I say it this way: I want to join 24 divided by 4 with 24 divided by 6 that you see makes 10. Therefore put 24 divided by 4 in the manner of a fraction which comes out to $\frac{24}{4}$. Again similarly, put 24 divided by 6 in the manner of a fraction. Then you must cross-multiply, which is 24 by 6 making 144 and now multiply 4 by 24 which is over the 6. It makes 96, joined with 144 making 240. Now multiply what is under the line, which are 4 and 6, making 24. Then you must divide 240 by 24, which comes to 10. I say that if

I multiply 10, which it must come to, against the divisor 24, that I make the quantity multiplied, that is 240, and so it is correct. Now we return to our problem. We take 100 divided by a thing and 100 divided by a thing plus 5 and so put these two parts as if they were a fraction in the way you saw drawn before. And now cross-multiply, the way we did before, which is 100 by a thing which makes 100 things. Now multiply the other split, which is 100 by a thing and 5, making 100 things and 500 numbers. Joined with 100 things, you have 200 things plus 500 numbers. Now multiply that which you have under the line against the other, which is one thing by one thing and 5, making a censo and plus 5 things. Now multiply the events, which are 20 against a censo and plus 5 things, making 20 censi and 100 things,

100	100
by a thing	by a thing plus 5

which quantity is equal to 200 things and 500 numbers. Now remove 100 things from each part. You will have that 20 censi are equal to 100 things and 500 numbers. Bring it to one censo, which is that you divide each part by the censi. You will have that one censo is equal to 5 things and 25 numbers. Halve the things, it comes to $2\frac{1}{2}$, multiply it by itself making $6\frac{1}{4}$, join it to the number, which is 25, making $31\frac{1}{4}$ and we say that the value of the thing is the root of $31\frac{1}{4}$ plus the half which is $2\frac{1}{2}$. And we supposed that the first divisor was a thing, therefore you first divide 100 by the root of $31\frac{1}{4}$ plus $2\frac{1}{2}$ and then divide by 5 more, so you divide by the root of $31\frac{1}{4}$ plus $7\frac{1}{2}$ and it is done.

14. Someone divides 100 by a quantity and then divides 100 by 5 more than the first, and that which each division comes to, subtracted one from the other, leaves 10. I want to know by what 100 was divided first and by what it was divided afterwards. Suppose that you divide 100 by a thing, it comes to 100 divided by a thing, and then you want to divide by 5 more, which is a thing and 5, which comes to 100 divided by a thing and 5. Now you must take one from the other and it must come to 10. Now cross-multiply, the way you did it before, which is 100 by a thing

making 100 things. Now multiply the other split, which is 100 by a thing and 5, making 100 things and 500 numbers. Now take away 100 things from 100 things plus 500 numbers and remain with 500 numbers. Now multiply what you have under the line, one against the other, which is a thing by a thing and 5 making a censo and 5 things. Then you must divide 500 numbers by a censo and 5 things and it must come to 10. So multiply one censo and 5 things by 10, making 10 censi and 50 things and this is equal to 500. Divide by the censi. You will have that one censo and 5 things are equal to 50 numbers. Halve the things, which are 5, it comes to $2\frac{1}{2}$, multiply by itself, making $6\frac{1}{4}$, join it to the number which is 50, you will have $56\frac{1}{4}$ and say that the thing is worth the root of $56\frac{1}{4}$ less the half of the thing, which was $2\frac{1}{2}$. And we supposed that you divided 100 first by a thing, so you first divided 100 by the root of $56\frac{1}{4}$ less $2\frac{1}{2}$ and then you divided by 5 more and so put 5 onto the root of $56\frac{1}{4}$ less $2\frac{1}{2}$. You will have the root of $56\frac{1}{4}$ plus $2\frac{1}{2}$ and say: the root of $56\frac{1}{4}$ plus $2\frac{1}{2}$ divides [the second] and then it is done.

15. Someone divided 100 by a quantity and that which it came to, multiplied by itself, made the divisor. I want to know by what 100 was divided. Suppose that you divide 100 by a thing. It comes to 100 divided by a thing. Now multiply 100 divided by a thing by itself, making 10000 divided by a censo. Then you must divide 10000 by a censo and it must come to the divisor which is the thing we supposed to be the divisor. So multiplying a thing by a censo makes a cube that is equal to 10000. Now you must divide 10000 by one cube. It comes to 10000 and I will say that the thing is worth the cube root of 10000 and do similar problems this way.

16. Someone divides 100 by a quantity and then divides 100 by 5 more than the first and multiplying one event by the other, makes 50. I want to know how much 100 was divided by first and by how much afterwards. Suppose that you divide 100 by a thing. It comes to 100 divided by a thing and put it as fraction. Then say that 100 is divided by 5 more than the first so you must divide 100 by a thing and 5 which comes to 100 divided by a thing and 5 and also put it as a fraction, as you saw drawn

before. Now you must multiply 100 divided by a thing against 100 divided by a thing and 5. And you are well advised that if you had to multiply $\frac{2}{5}$ by $\frac{3}{7}$ it would do you well to do it this way, that is, you multiply the 2 which is over the 5 against the 3 which is over the seven, which make vi [6]. Then you multiply the 5 which is under the line against the 7 which is also under the line, which makes 35. Now you divide 6 by 35 which comes to $\frac{6}{35}$ and this is how much multiplying $\frac{2}{5}$ by $\frac{3}{7}$ comes to. Therefore in our problem, you must multiply the 100 divided by a thing against 100 divided by a thing and 5. So you must multiply the 100 that is over the thing against the 100 that is over the thing and 5 which makes 10000. Now multiply a thing by a thing and 5, making a censo and 5 things. Then you must divide 10000 by a censo and 5 things and it must come to 50. So if you multiply 50 by a censo and 5 things, that which it makes will be equal 10000. Then multiply 50 by a censo and 5 things, making 50 censi and 250 things.

100	100
by a thing	by a thing and 5

And this way we have that 50 censi and 250 things are equal to 10000 numbers. Divide by the censi. You will have that one censo and 5 things are equal to 200 numbers. Halve the thing, which is 5, it comes to $2\frac{1}{2}$. Multiply it by itself, making $6\frac{1}{4}$, join it to the number which is 200. You will have $206\frac{1}{4}$ and say that the thing is worth the root of $206\frac{1}{4}$ less the half which is $2\frac{1}{2}$. And we supposed that we divided 100 by a thing, so you first divide 100 by the root of $206\frac{1}{4}$ less $2\frac{1}{2}$ and then you divide by 5 more. Put it onto the root of $206\frac{1}{4}$ less $2\frac{1}{2}$ to get that you divide by the root of $206\frac{1}{4}$ plus $2\frac{1}{2}$ and then it is done.

17. Someone divided 100 by a quantity and that which it came to, put onto the divisor, made 30. I want to know by what 100 was divided. Suppose that you divided 100 by a thing. Then you must join 100 divided by a thing to the divisor which is a thing and it must make 30. Now I must present a clear example which will advise you well and I want to say it this way: I want to divide 40 by 5 and that which it comes to, put onto 12, makes

20. I say that if I take the 12 that I said before from 20, that which remains will be as much as the event of 40 divided by 5. Then you must apply this same way to our problem, which is that you must join 100 divided by a thing with a thing and it must make 30. Then if I take a thing from 30, that which remains will be as much as the event of 100 divided by a thing. So take a thing from 30 and remain with 30 less a thing. Then dividing 100 by a thing must come to 30 less a thing. So if I multiply the divisor which is a thing, against that which it must come to, which is against 30 less a thing, that which it makes is the quantity that you want to divide, which is 100. So you must multiply 30 less a thing against a thing, which make 30 things less a censo and this way we say that 30 things less a censo is equal to 100 numbers. Restore the parts, which is that you must give a censo to each part. You will have that 30 things are equal to a censo and 100 numbers. You must divide by the censi, which is by one, leaving it the same. Halve the things which are 30. The half is 15. Multiply it by itself making 225, subtract the number which is 100, remain with 125 and I say that the thing is worth the root of 125 plus the half of the things which was 15. And we supposed that 100 was divided by a thing, therefore you divided 100 by the root of 125 plus 15. And you must do similar ones this way.

18. Someone divided 100 by a quantity and that which it came to, subtracted from the divisor, made to 20. I want to know by what 100 is divided. Suppose that 100 is divided by a thing. It comes to 100 divided by a thing. Then you must take 100 divided by a thing from a thing. What remains must be 20. And so as to be well advised for this problem, I want to pose a clearer example and I want to say it this way: I want to divide 40 by 10 and take away that which it comes to from the divisor which is 10. There remains 6. I say that if you take 6, which must remain from 10, there remains the event of 40 divided by 10 and take 6 from 10, remain with 4. And dividing 40 by 10 certainly comes to 4 and so you will be certain. Therefore, by this same consideration, do our problem. And we have to take 100 divided by a thing from a thing and must remain with 20. So if you take 20 from the divisor which is a thing, I say that

what remains will be as much as the result of 100 divided by a thing. Therefore, take 20 from a thing. Remain with a thing less 20 numbers. Then the event of 100 divided by a thing must be a thing less 20. So multiply a thing less 20 by the divisor which is a thing and that which it makes will be equal to 100. Then multiply a thing less 20 by a thing making a censo less 20 things and this is equal to 100. Restore the parts, which is that you must give 20 things to each part. You will have that one censo is equal to 20 things and 100 numbers. You must divide by the censi, which comes to the same. Halve the things which are 20, the half is 10, multiply it by itself, making 100, joined to the number which is 100 and have 200. And we say that the value of the thing is the root of 200 plus the half of the things which was 10 and it is done.

19. Someone divided 100 by a quantity and what it came to, multiplied against the root of the divisor, made 25. I want to know what 100 is divided by. Suppose that you divide 100 by a thing. It comes to 100 divided by a thing. Then you must multiply 100 divided by a thing against the root of a thing. So you must bring 100 divided by a thing to the root which makes 10000 divided by a censo. You must multiply the root of 10000 divided by a censo against the root of a thing, which makes the root of 10000 things divided by the root of a censo. Then you must divide the root of 10000 things by the root of a censo and it must come to 25. So you must bring 25 to the root which makes 625. Now multiply the root of a censo against the root of 625 which makes the root of 625 censi and you will say that 625 censi are equal to 10000 things. You must divide 10000 by 625, which comes to 16 and we say that that the thing is worth 16. And we supposed that 100 was divided by a thing. So you divide 100 by 16 and it is done.

20. Someone divided 100 by a quantity and then divided 100 by 5 more than the first and then divided it by twice as much as [went into] the first, and these three events joined together make 20. I want to know how much 100 is divided by each time. Suppose that you divide 100 by a thing. It comes to 100 divided by a thing. Then say that it is divided by 5 more than the first.

Therefore you divide it by a thing plus 5 numbers which comes to 100 divided by a thing plus 5. Then you divide by twice as much as the first, which is by 2 things. It comes to 100 divided by two things. Put it in the order that you saw here before, like 3 fractions. Now you must join these three events and they must make 20. And continue in this way, which is that you first join 100 divided by a thing with 100 divided by a thing plus 5. Cross-multiply this way which is 100 and by a thing and 5, making 100 things and 500 numbers. Now multiply the other split, which is 100 by a thing which makes 100 things, joined to 100 things and 500 numbers, makes 200 things plus 500 numbers. Now multiply that which is under the line, one against the other, which is a thing by a thing and 5, which make a censo and 5 things.

And so you have joined together the first two events and they make 200 things plus 500 numbers divided by a censo and 5 things. Now with this you must join the third event

$$\frac{100}{\text{by a thing}} \quad \frac{100}{\text{by a thing and 5}} \quad \frac{100}{\text{by 2 things}} \quad \frac{\text{200 things and 500 numbers}}{\text{by a censo and 5 things}} \quad \frac{100}{\text{by 2 things}}$$

which is 100 divided by two things. And similarly put them as you did below[19] like two fractions. So that you are better advised, put them like you see next[20] which is 200 things plus 500 numbers divided by a censo and 5 things. Also put the third event which is 100 divided by 2 things. Now cross-multiply which is 100 by a censo and 5 things, they make 100 censi and 500 things. Now multiply by the other split which is 2 things by 200 things and 500 numbers, they make 400 censi and 1000 things, joined with 100 censi and 500 things, make 500 censi and 1500 things. Now multiply that which is under the line, one against the other, which is one censo and 5 things by 2 things, making 2 cubes and 10 censi. So you have joined the three events and you have that there are 500 censi and 1500 things divided by 2 cubes and 10 censi. Now you must divide 500 censi and 1500 things by 2 cubes and 10 censi and they must

[19]The diagram is placed here exactly as it is positioned in the manuscript. It may have been differently placed in the scribe's source.

[20]See previous note.

come to the sum of the events which is 20. So multiply 20 by two cubes and 10 censi. They make 40 cubes and 200 censi and this is equal to 500 censi and 1500 things. Now you must restore the parts which is that you take 200 censi from each part. You will have that 40 cubes are equal to 300 censi and 1500 things. You must divide by the cubes, which is by 40. You will have that one cube is equal to $7\frac{1}{2}$ censi and $37\frac{1}{2}$ things.[21] Halve the censi, which are $7\frac{1}{2}$, the half is $3\frac{3}{4}$, multiplied by itself, making $14\frac{1}{16}$, join it to the things which are $37\frac{1}{2}$, making $51\frac{9}{16}$ and say that the thing is worth the root of $51\frac{9}{16}$ plus the half of the censi, which was $3\frac{3}{4}$. And we supposed that 100 was first divided by a thing, therefore you divide the first by $3\frac{3}{4}$ plus the root of $51\frac{9}{16}$. And then you divide by 5 more. Thus the second time you divide by $8\frac{3}{4}$ plus the root of $51\frac{9}{16}$. The third time, you divide by twice the first, which is $7\frac{1}{2}$ plus the root of $206\frac{1}{4}$. I want to say why $206\frac{1}{4}$: because you know that the root is multiplied by 4 and it is done.

21. Someone lends to another 100 lire at the rate of some denari[22] per lira per month. And when the end of the year comes and the one who has these 100 lire on loan wants to return the capital and the interest to the one who lent them to him, the one who lent this money says: Keep it for another year at the rate of two [denari] per lira more per month than before and when the second year comes he returns 120 lire to him. I ask at what rate per month the lira was lent the first year. We must do it this way: suppose that one lira was lent at one thing per month, which per year is worth 12 things, which for the 12 things gets $\frac{1}{20}$ of 100 lira which is 5 things, that added to 100 lire, make 100 lire and 5 things. And then say that for the second year the lira earns two more denari per month than in the first.

    Now we know that in the next year, the lira per month is worth two more denari per lira per month which is worth 12 things and 2 soldi per year. We must compute the interest on 100 lire and 5 things which in truth earns 10 and $5\frac{1}{2}$ things and $\frac{1}{4}$

---

[21]This cubic is solved correctly by Rule 10.

[22]Currency conversions: 1 lira = 240 denari , 1 lira = 20 soldi, 1 soldi = 12 denari.

censo.[23] Joined to the 100 lire and 5 things, you have 110 lire and $10\frac{1}{2}$ things and $\frac{1}{4}$ censo and all of this is equal to that which is returned, which is 120 lire. And so we have that 110 lire and $10\frac{1}{2}$ things and $\frac{1}{4}$ censo are equal to 120 lire.

Take 110 lire from each part. You will have that $10\frac{1}{2}$ things and $\frac{1}{4}$ censo are equal to 10 lire. You must divide by the censi which is by $\frac{1}{4}$ and you will have that 42 things and a censo are equal to 40. Halve the thing which is 42, the half of which is 21, multiply it by itself, which is 441, join it with the number which is 40. You will have 481. We have that the thing is worth the root of 481 denari less the half which was 21. And we suppose that the a lira was lent at a thing, therefore was lent per lire per month at the root of 481 denari less 21 in the first year and it is done.

22. Someone lends another 30 lire for 2 years to be done [compounded] at the end of the year. When it comes to the end of the two years, he returns 30 and the root of 30 lire. I want to know at what rate the lira was lent per month. Always keep this rule in mind, that in one year it comes by number, and in two years it always comes by root, and in 3 years it comes by cube root, and in 4 years it comes by root of root, and in 5 years it comes by root [by] cube root and in 6 years it comes by [cube] root [by] [cube] root and it goes on, graded by whatever I say.[24] And this problem says two years and so it comes by root. Now I want to teach you another rule that you always use in these problems: 20 is always brought to the roots that are in years, that is, to the roots or the cube roots and this way for other roots.[25] Take 20 to the root, which is 20 by 20, making 400. Also I want to teach you another thing, that you always divide the capital and the interest by the capital which is in our problem that you want to divide 30 and the root of 30 by 30.

[23]The interest rate the second year is $x+2$ denari per month $= 12x+24$ denari per year $= (12x+24)/240$ lire per year $= x/20+1/10$.

[24]If an amount A is lent a rate of r per year then after one year multiply A by the number $(1+r)$, after two years multiply by $(1+r)^2$ (i.e, by what $(1+r)$ is the root of), after three years by $(1+r)^3$, etc.)

[25]The problem now is to get the rate in denari per month. 1 lira per year is 240 denari per year which is 20 denari per month. Here is the algebra: Per year: $30(1+x)^2 = 30 + (30)^{1/2}$ so $(1+x)^2 = 1 + (1/30)^{1/2}$. Per month in denari: $x = 20((1+(1/30)^{1/2})^{1/2} - 20$. Note that this problem is not done with algebra.

In truth it comes to one and the root of $\frac{1}{30}$. And then always multiply the 400 that you took to the root against one and the root of $\frac{1}{30}$, so that 400 by 1 makes 400. And now multiply 400 by the root of $\frac{1}{30}$. You must bring 400 to the root, which makes 160000. Now multiply 160000 by $\frac{1}{30}$ making the root of $5333\frac{1}{3}$. And you respond that the lira was lent per month at the root of $5333\frac{1}{3}$ put onto 400. The root of that sum less 20 denari [is the rate] and it is done.

23. Someone lends another 100 lire and 100 gold florins for a year and I do not know at what rate it is lent. When it comes to the end of a year, the one who borrowed returned 10 lire and 2 florins from the 100 lire and 10 florins and 5 lire from the 100 florins. I want to know how much the florin is worth [in lire]. Suppose that the florin is worth one thing of lire to him. Now you know how much 100 florins are worth, which is 100 things. From those 100 things, he returns 10 florins and 5 lire. You know that the 10 florins are worth 10 things at 1 thing each. So we have found that he lent 100 things and he got 10 things and 5 lire in interest. Now we know that he gave him 10 lire and two florins in interest from the 100 lire. We know how much two florins are worth at one thing each, that they are worth two things. Therefore from the 100 lire that he lent, he takes in return 10 lire and two things. Now multiply 100 lire by 10 things and 5 lire, making 1000 things and 500 lire. Now multiply the other split, which is 100 things by 2 things and 10 lire, making 200 censi and 1000 things and so that you know this multiplication, I draw it here.

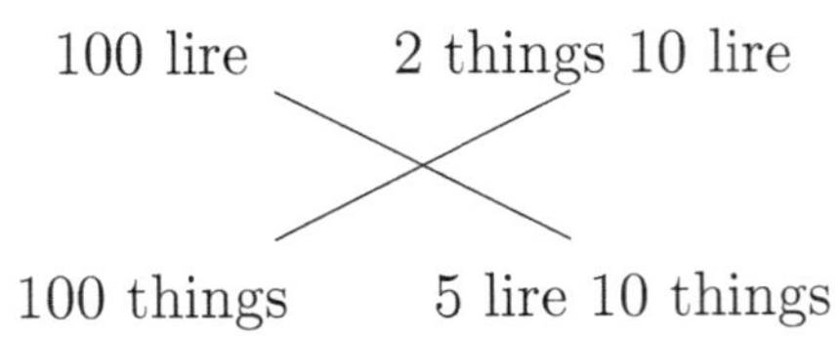

Now say that one thousand things and 500 lire are equal to 200 censi and 1000 lire[26]. Now you must restore the parts, which is

[26]The scribe erred. Lire should be things. The mistake continues so the rest of the algebra does not solve the problem.

that you take 500 lire from each part. You will have that 1000 things are equal to 200 censi and 500 lire. You must divide by the censi, which are 200. You will have that 5 things are equal to one censo and $2\frac{1}{2}$ lire. Halve the things, which are 5. The half is $2\frac{1}{2}$. Multiply it by itself, making $6\frac{1}{4}$, subtract the number which is $2\frac{1}{2}$, remain with $3\frac{3}{4}$. And we supposed that the florin is worth a thing. Therefore the florin is worth $2\frac{1}{2}$ plus the root of $3\frac{3}{4}$ and it is done.

24. Someone wants to exchange 100 florins for piccioli and fiorini. The florins are worth twice as much in piccioli than in fiorini less 6 piccioli. Now he exchanged these 100 florins and he got 100 fiorini and 100 piccioli. I ask what the florins are worth in fiorini and what in piccioli. We must do it this way. We suppose that the florins are worth one thing in fiorini and we deduce that they are worth twice as much in piccioli as in fiorini. Therefore they were worth two things less 6 soldi in piccioli. Now I want to tell it to you this way: if a florin is worth 3 lire and you want to know how many florins are in 100 lire, you divide 100 lire by 3 and what it comes to would be the florins that you have. And you do the fiorini in a similar way. Now you had supposed that the florin is worth a thing of fiorini. Therefore you must divide the 100 lire of fiorini that he got by a thing, which comes to 100 divided by a thing of fiorini. Now you must divide the 100 lire of piccioli by 2 things less 6 soldi. It comes to 100 divided by 2 things less 6 soldi in piccioli and put 100 divided by a thing as a fraction and also similarly for 100 divided by 2 things less 6 soldi in piccioli. Now you must join these two, which are 100 divided by a thing and 100 divided by 2 things less 6 soldi of piccioli. And join them in this way, which is as if you wanted to join $\frac{1}{3}$ and $\frac{1}{4}$ which is that you cross multiply, which is multiply 4 by 1 making 4 and now multiply iii by one, making 3, joined together making 7. And this you want to

100	100
by a thing	by two things less 6

divide by that which is under the line, multiplied one against the other, which is 3 by 4 making 12. Divide 7 by 12, it comes to $\frac{7}{12}$. Now you must do the above this way, which is that you multiply similarly, which is that you must multiply 100 by 2 things less 6, which makes 200 things less 600. Now multiply the other cross, which is 100 by a thing, making 100 things. Join together like you did above, making 300 things less 600 numbers. And you want to divide this by that which is under the line, multiplied one against the other, which is one thing by 2 things less 6, making 2 censi less 6 things. And so you must divide 300 things less 600 numbers by 2 censi less 6 things and it must come to the 100 florins that he exchanged at first. You must multiply 100 by 2 censi less 6 things by making 200 censi less 600 things and this is equal to 300 things less 600 numbers. Restore each part, to each part give 600 numbers. You will have that 300 things are equal to 200 censi [less] 600 things plus 600 numbers. Again you must restore each part, which is that you must give 600 things to each part. You will have that 900 things are equal to 200 censi plus 600 numbers. This is the fifth rule which is that you must divide by the censi. You will have that $4\frac{1}{2}$ things are equal to a censo plus 3 numbers. Now you must halve the things which are $4\frac{1}{2}$, the half is $2\frac{1}{4}$, multiply by itself, making $5\frac{1}{16}$, remove the number which is 3, remain with $2\frac{1}{16}$ and say that the thing is worth the half which was $2\frac{1}{4}$ plus the root of $2\frac{1}{16}$. And we supposed that the florins were worth a thing per fiorini, therefore a florin is worth $2\frac{1}{4}$ plus the root of $2\frac{1}{16}$ fiorini and we supposed that, for the piccioli, 2 things less 6 numbers. Multiply 2 by $2\frac{1}{4}$ plus the root of $2\frac{1}{16}$, making $4\frac{1}{2}$ plus the root of $8\frac{1}{4}$. Remove 6 that the value lacks, remain with the root of $8\frac{1}{4}$ less $1\frac{1}{2}$ and the worth per piccioli is the root of $8\frac{1}{4}$ less one and $\frac{1}{2}$ and it is done.

25. A merchant must have from another, who is a tavoliere [a money changer], 100 lire in fiorini and when it comes to the end of some period of time and this merchant returns to the tavoliere and says, "I want the exchange of those 100 lire in fiorini and write them as piccioli." And the tavoliere does so and gives him in the exchange for those 100 lire in fiorini and then he wrote that he must have 100 lire in piccioli. And when it came to the end of another period of time and this merchant returns to the tavoliere and says, "Those 100 lire that you wrote for me in

piccioli, I want you to write them for me in fiorini." And the tavoliere does so and the merchant says how many lire in fiorini they are and he says, "They are as many as I gave you for the exchange of 100 lire in fiorini at first." I ask for the lira in fiorini that he received from the exchange of a lira of piccioli.

Suppose that he [the tavoliere] receives a thing for an exchange, therefore the 1 lira of fiorini is worth 1 lira and 1 thing of piccioli. Now we know what the 100 lire of fiorini are worth, that they are worth 100 lire and 100 things of piccioli and the tovoliere wrote 100 lire of piccioli. So in the exchange what he made was 100 things and keep in mind 100 things.

Now he says that he wants to know how much the 100 lire of piccioli that he wrote him are worth in fiorini. Therefore you must divide the 100 lire by 1 and 1 thing and I say that this must come to the 100 things that he got for the exchange. Then multiply one lire and one thing by 100 things, making 100 things and 100 censi and this is equal to 100 lire. Divide by the censi and you have that one thing and one censo are equal to 1 lire. Halve the thing, which is $\frac{1}{2}$, multiply it by itself, making $\frac{1}{4}$, and join it to the number. You will have $1\frac{1}{4}$ and you say that the the thing is worth the root of one and $\frac{1}{4}$ less $\frac{1}{2}$.

And we supposed that, for the exchange of a lira in piccioli for a lira in fiorini, he gets one thing. Therefore he receives the root of one and $\frac{1}{4}$ less $\frac{1}{2}$ in fiorini from that piccioli which could be solde xii, denari vi and indeed anything else.

26. Someone has 12 florins and another has 50 aquilini and the one that has 12 florins wants to change to aquilini and the one that has aquilini wants to change to florins. They exchanged one into the other at the same rate and when they had exchanged this way, the one who had 50 aquilini finds he has as many florins that, subtracted from the aquilini of the one who had 12 florins, leaves 6. I want to know how much the florin is worth in aquilini. Suppose that the florin is worth one thing in aquilini. Therefore, the 12 florins are worth 12 things. And then say that subtracting the florins that he had from those 50 aquilini, there remains 6. I want to give you this example and I want to say that the florin is worth 5 aquilini, so the 12 florins will be worth 60 aquilini. Now

you know the florin to be worth 5 aquilini. When the florins are divided into 50 aquilini and the florins that you divided are subtracted from 60, there must remain 6.

Now you supposed that the florin is worth one thing of aquilini, therefore the 12 florins are worth 12 things of aquilini like we said before. Now we know how many florins are in 50 aquilini at one thing per florin, which is that you must divide 50 by one thing, that comes to 50 divided by a thing. And these are the florins that he got from the 50 aquilini and you must subtract these florins from the 12 things, remaining with 12 things less 50 divided by one thing. And we said before that subtracting the florins that he had from the 50 aquilini from the aquilini that he had from the 12 florins, there remains 6. Therefore 12 things less 50 divided by a thing are equal to 6.

And so bring each thing to the numerator,[27] which is that you bring to the whole by a thing and multiply a thing by 12 things making 12 censi, joined to the subtracted 50 and you have 12 censi less 50 numbers. Now you must multiply a thing by 6, making 6 things. Restore the parts. Give each part 50 numbers to get that 12 censi are equal to 6 things and 50 numbers. Divide by the censi, which is by 12, you will have that one censo will be equal to $\frac{1}{2}$ thing plus $4\frac{1}{6}$ numbers. Halve the things, get $\frac{1}{4}$, multiply it by itself, making $\frac{1}{16}$, join it to the number which is $4\frac{1}{6}$. You will have 4 and $\frac{11}{48}$ and respond that the value of a thing is the root of 4 and $\frac{11}{48}$ plus the half of the things which was $\frac{1}{4}$. And we supposed that the florin is worth one thing per aquilino and do similar problems this way.

27. Someone has 12 florins and someone else has 20 lire and he who has the florins wants to change them to lire and he who has the lire wants to change them to florins. And they changed one to the other at the same rate. When they had changed this way, between the lire that he had from 12 florins and the florins that he had from 20 lire, joined together, make 64. I want to know how many lire the florin is worth. Suppose that the florin is worth one thing of lire. Now you must change the 20 lire to florins, at one thing per one [florin] so you must divide the 20

[27]The author is multiplying through by the "thing" in the denominator.

lire by a thing which comes to 20 divided by a thing. Now you know that the 12 florins, at one thing per one [florin], are worth 12 things. Joined with the lire that he got from the florins, which is with 20 divided by a thing, they make 12 things plus 20 divided by one thing, and these joined must make 64. So we have 12 things plus 20 divided by one thing are equal to 64. Bring it to a numerator, which is that you bring to the whole by one thing, which is that you multiply a thing by 12 things and you join 20 making 12 censi plus 20 numbers. Now multiply a thing by 64, making 64 things. Thus we have that 12 censi and 20 numbers are equal to 64 things. Divide by the censi. You will have that $5\frac{1}{3}$ things are equal a censo and to one and $\frac{2}{3}$ numbers. Halve the things, getting 2 and $\frac{2}{3}$, multiply it by itself, making 7 and $\frac{1}{9}$, take away 1 and $\frac{2}{3}$, remain with 5 and $\frac{4}{9}$. And say that the thing is worth the root of 5 and $\frac{4}{9}$ plus 2 and $\frac{2}{3}$ which is the half of the things. And we supposed that the florin is worth one thing. Therefore the florin is worth the root of $5\frac{4}{9}$ plus 2 and $\frac{2}{3}$ and it is done.

28. Someone has 10 florins and another has 60 viniziani and the one that has florins wants to change to viniziani and the one that has viniziani wants to change to florins. And they change one into the other at the same rate. And when they exchanged this way, the one who had 60 viniziani finds as many florins that, multiplying it by itself, makes as many as were the viniziani that the one had from the 10 florins. I want to know what the florin is worth in viniziani. Because the florin is worth one thing of viniziani, the 10 florins are thus worth 10 things of viniziani. Now you should know how many florins are in 60 viniziani at one thing per florin, for which you must divide 60 by a thing. So you must multiply 60 divided by a thing by itself which makes 3600 divided by a censo and they must make 10 things. Then multiply 10 things against one censo, making 10 cubes. Then we have that 10 cubes are equal to 3600. You must divide 3600 by the cubes, which are 10. It comes to 360. We say that the thing is worth the cube root of 360 viniziani. And we had supposed that the florin is worth one thing, therefore the florin is worth the cube root of 360 viniziani and it is done.

29. Someone has a florin that can get 15 grossi and 100 denari. And the florin is worth as many grossi as the grosso is worth in denari. I want to know how many grossi the florin is worth and how many denari the grosso is worth.[28] Always do it this way, which is that you halve the grossi that he got from the florins, which is 15. The half is $7\frac{1}{2}$ and always multiply this by itself, which makes $56\frac{1}{4}$. And you always want join onto the quantity of denari, which is 100. You will have $156\frac{1}{4}$ and say that the florin is worth the root of $156\frac{1}{4}$ plus the half of the grossi, which is $7\frac{1}{2}$. And the grosso is worth the root of $156\frac{1}{4}$ plus $7\frac{1}{2}$ denari and it is done.

30. Someone has a florin for which he can get a quantity of tornesi. And the tornese is worth twice as many in denari di piccioli as the florin is worth in tornesi. Now he exchanged this florin and he got 8 tornesi and 8 soldi di piccioli. I want to know how much the florin is worth in tornesi and how much the tornese is worth in denari piccioli. Suppose that the florin is worth 1 thing of tornesi. Then it must be that the tornese is worth twice as much as a thing, which is two things. Therefore the tornese is worth two things in denari di piccioli. Now you must find out how much the florin is worth in piccioli. I say if the florin is worth one thing of tornesi [and] that, if you know what the value of a tornese per thing of piccioli is, the florin is worth that much in piccioli. Therefore to know what one thing of tornesi is worth at 2 things for one, you must multiply the quantity that the tornesi are worth against the quantity of tornesi. Therefore you must multiply two things by one thing, making 2 censi. So the florin is worth two censi in piccioli. To find out how much each coin is worth, divide viii tornesi by a thing that we had supposed was the value of the florin, which comes to viii tornesi divided by a thing and also divide 8 soldi, which are 96 denari, by two censi which we found that the florin was worth in piccioli, which comes to 96 divided by 2 censi. And put these two divisions as two fractions like you see here below and the results of these two divisions, joined together, must make one florin. And then,

[28]The scribe does not set up the algebra problem. Let the florin be worth x grossi and then the grosso is worth x denari. So the florin is worth x^2 denari and $x^2 = 100 + 15x$ which is the equation he solves.

when you have put them in order this way, you cross multiply, which is 8 by two censi, making 16 censi. And then multiply 96 by a thing, making 96 things, joined with 16 censi, you have 16 censi and 96 things. Now multiply that which is under the line, one against the other,

tornesi 8	/	denari 96
by a thing	/	by two censi

which is one thing by 2 censi, which make 2 cubes. Therefore we have brought the two divisions into one and we have that 16 censi and 96 things are divided by 2 cubes and it must come to one. Therefore multiply one by 2 cubes, making 2 cubes. We have that two cubes are equal to 16 censi and 96 things. Divide by the cubes, and we have that one cube is equal to 8 censi and 48 things You must halve the censi which are 8, the half is 4, multiply it by itself, making 16, join it to the things, which are 48, making 64 and we say the thing is worth the root of 64 plus the half of the censi which was 4. And we had supposed that the florin was is worth one thing, therefore the florin is worth the root of 64 plus 4 tornesi in tornese, and the tornese is worth twice the root of 64 plus 4 which make the root of 256 plus 8. The root of 64 is 8, plus 4, and you have 12. Therefore it is worth 12 in tornesi. Now we take the root of 256, which is 16, joined with 8. You have 24 and the tornese is worth 24 in piccioli denari which is certainly twice as much as 12 tornesi and it is done.

31. Someone has one florin from which he can get a quantity of tornesi and further, he wants twice as many vinizian as tornesi. Now he exchanged them and he got 8 tornesi and 8 vinizian. I want to know what the florin is worth in tornesi and what in vinizian. Suppose that the florin was worth a thing in tornesi which in turn was worth twice as much as a thing in vinizian so would be worth 2 things. Next you must divide 8 tornesi by one thing and 8 vinizian by two things and to join them together you must put them in order like you see here next. And then cross multiply, which is a thing by 8, making 8 things, and then multiply 8 by 2 things, making 16 things, joined together,

makes 24 things. Next, multiply that which is under the line one against the other, which is one thing by two things, making 2 censi. Therefore you must divide 24 things by 2 censi and this must come to one. You must multiply 2 censi by one, making 2 censi. We have that two censi are equal to 24 things. So you must divide 24 things by 2 censi. It comes to 12. We say that the thing is worth 12. And we had supposed that the florin is worth a thing of tornesi. Therefore the florin is worth 12 tornesi and the vinizian is worth twice as much as 12, so the vinizian is worth 24 and it is done.

8	/	8
by a thing		by two things

32. Someone has one florin that can get a quantity of tornesi and further he wants three times as many plus 3 viniziani. And he exchanged it and got 10 tornesi and 10 viniziani. I want to know the value of the florin in tornesi and in viniziani. Suppose that a florin is worth a thing in tornesi, which makes the value in viniziani three times as much plus 3. So in vinizi, it would be worth 3 things plus 3 viniziani. Next, you must divide 10 tornesi by a thing and 10 viniziani is divided by a thing[29] plus 3.

10	/	10
by a thing		by a thing and 3

Put it in the order that you did before and then cross multiply which is 10 by 3 things plus 3, making 30 things plus 30 numbers. Now multiply the other split, which is 10 by a thing, making 10 things, joined with 30 things and 30 numbers, making 40 things plus 30 numbers. Now multiply that which is under the line, one against the other, which is a thing by 3 things and 3, making 3 censi and 3 things. Then you must divide 40 things plus 30 numbers by 3 censi and 3 things and it must come to one. So multiply one by 3 censi and 3 things, making 3 censi and 3 things. We have that 3 censi and 3 things are equal to 40 things

[29]It should be 3 things. The mistake occurs in the diagram as well. But the calculation that ensues is correct.

and 30 numbers. Now you must equalize the parts, which is that you take 3 things from each part. You will have that 3 censi are equal to 37 things and 30 numbers. Divide by the censi, which are 3, and you will have that one censo is equal to $12\frac{1}{3}$ things and 10 numbers. Halve the things, which are $12\frac{1}{3}$, the half is $6\frac{1}{6}$, multiplied by itself, is 38 and $\frac{1}{36}$, join it to the number, which is 10, making 48 and $\frac{1}{36}$ and we say that a thing is worth the root of 48 and $\frac{1}{36}$ plus the half of the things which was $6\frac{1}{6}$. And we supposed that the florin was worth a thing in tornesi, therefore it is worth the root of 48 and $\frac{1}{36}$ plus $6\frac{1}{6}$ in tornese and in vinizi, it is worth the three times as much plus 3. To know how much it is worth in vinizian you must bring the 3 to the root which makes 9, multiplied against the root of $48\frac{1}{6}$, makes $433\frac{1}{4}$. Now multiply 3 by $6\frac{1}{6}$, making $18\frac{1}{2}$, plus 3, [which is] $21\frac{1}{2}$. And we say the value in viniziani is the root of $433\frac{1}{4}$ plus $21\frac{1}{2}$ and in tornesi, it is worth the root of $48\frac{1}{36}$ and $6\frac{1}{6}$ and it is done.

33. Someone has a florin for which he can get a quantity of tornesi and further he wants twice as many viniziani plus two and further he wants 4 times as many genovini as tornesi. Now he exchanged it and he got 6 viniziani, 6 tornesi, and 6 genovini. I want to know in this same problem, how much the florin is worth in each coin. Suppose that the tornesi is worth a thing which in turn would be worth in viniziani twice as much plus 2 so 2 things plus 2 in viniziani and in genovini, 4 times as many as tornesi, and so it would be worth 4 things.

    Now you must know that 6 is divided by a thing and 6 is divided by 2 things plus 2 and that 6 is divided by 4 things. And these 3 results joined together must make 1. And put them in order like you see below.

6	/	6	/	6
by a thing		by 2 things plus 2		by 4 things

Then join together 6 divided [by a thing] and 6 divided by 2 things plus 2 and cross multiply which is 6 by a thing making 6 things. Now do the other split, which is 6 by 2 things plus 2 making 12 things plus 12, joined with 6 things making 18 things plus 12. Now multiply the one which is under the line against the other, which is one thing by two things plus 2 making 2 censi and 2 things. Thus we have joined the 2 first events and it makes 18 things plus 12 divided by two censi and 2 things. Now you must join the other event with this which is 6 divided by 4 things and put it in the way you did before and cross multiply, which is 6 by 2 censi and 2 things, making 12 censi and 12 things. Now multiply the other split, which is 18 things plus 12 by 4 things making 72 censi plus 48 things, joined with 12 censi and 12 things, you have 84 censi and 60 things. Now multiply that which is under the line, one against the other, which is 4 things by 2 censi and 2 things, making 8 cubes and 8 censi. Then you must divide 84 censi and 60 things by 8 cubes and 8 censi and it must come to 1 so multiply 1 by 8 cubes and 8 censi, making 8 cubes and 8 censi and this is equal to 84 censi and 60 things. Restore the parts. Take 8 censi from each part and you will have that 8 cubes are equal to 76 censi and 60 things. Divide by the cubes, which are 8. It comes to $9\frac{1}{2}$ censi and $7\frac{1}{2}$ things. Now halve the censi, which is $9\frac{1}{2}$, the half of which is $4\frac{3}{4}$ multiplied by itself making $22\frac{9}{16}$, joined to the things, making $30\frac{1}{16}$. And we say that the thing is worth the root of $30\frac{1}{16}$ plus the half of the censi which was $4\frac{3}{4}$. And we supposed that the florin is worth one thing in tornesi, therefore the florin is worth the root of $30\frac{1}{16}$ plus $4\frac{3}{4}$ in tornesi, and join 2 and that much you will have in genovini which you will find them indeed worth the root of $120\frac{1}{4}$ plus $11\frac{1}{2}$. And the genovini are worth 4 times as much as the tornesi. Thus they are worth the root of 481 plus 19. And this way we have that it is worth the root of $30\frac{1}{16}$ plus $4\frac{3}{4}$ in tornesi, and it is worth the root of $120\frac{1}{4}$ plus $11\frac{1}{2}$ in viniziani, and it is worth the root of 481 plus 19 in genovini and it is done.

34. A well-digger must dig a well that is 20 braccia[30] deep and he should get 20 lire. Now he came and the amount he dug got him xii [12] lire. I want to know how many braccia he dug. Reproducing the work braccio by braccio, do it this way: always collect one by one up to 20 which makes 210.[31] Now say: From

---

[30] A braccio (pl. braccia) was a unit of length approximately the length of an arm. In Florence, a braccio or "arm" was about 23 inches. It varied from region to region.

[31] The pay was in proportion to the depth dug, one unit of pay for the first

210 dug, if I got 20 lire, how much was dug that got 12 lire? Multiply 12 by 210, making 2520, divide by 20. It comes to 126. And you say that he dug, digging one by one, to that number which makes 126. To know how much that number was, now collect one by one up to a thing, then you must do this: take $\frac{1}{2}$ of a thing, which is $\frac{1}{2}$ thing. Now put one onto the thing, making one thing and one. Now multiply $\frac{1}{2}$ thing by one thing and one, making $\frac{1}{2}$ thing and $\frac{1}{2}$ censo, and $\frac{1}{2}$ thing and $\frac{1}{2}$ censo is equal to 126. Divide by the censi, which is $\frac{1}{2}$. You will have that one censo and one thing is equal to 252. Halve the thing, which is one, the half of which is a $\frac{1}{2}$, multiply by itself, making $\frac{1}{4}$, join it to the number, making $252\frac{1}{4}$. And we supposed that the number was a thing therefore he dug the well to the root of $252\frac{1}{4}$ less $\frac{1}{2}$ braccio.

35. There are two men who have denari. The first says to the second, "If you give me 16 of your denari I will have twice as much as you." The second says to the first, "If you give me such a part of yours as I have given you of mine, I would have three times as much as you." I want to know how much each had. I want to teach you this problem by a nice rule by which you can do all similar problems. You know that [when] something is said to be twice as much, it is as two thirds of the whole quantity and three times as much as the said quantity is $\frac{3}{4}$ of the whole quantity. Now know in what number one can find $\frac{2}{3}$ and $\frac{3}{4}$ which you find to be 12. Now take $\frac{2}{3}$ of 12 which is 8 and take $\frac{3}{4}$ of 12 which is 9, joined with 8, makes 17. Take away 12, in which you found the numbers, there remains 5 and 5 is our divisor. Now multiply 12 in which you find the numbers, against what was asked for by the first from the second, which is 16, making 192, divided by 5, which you kept, it comes to $38\frac{2}{5}$ and $38\frac{2}{5}$ is what the second had. Now to know what the first had, do it this way. The first asked the second for 16 denari. Take 16 away from $38\frac{2}{5}$ remaining with $22\frac{2}{5}$. So the second remained with $22\frac{2}{5}$. And the first said he had twice as much as the second so it must be that he had $44\frac{4}{5}$ so you know that the first had $44\frac{4}{5}$ with the 16 that the second gave him. Therefore take 16 from $44\frac{4}{5}$ and

---

braccia dug, two for the second one down, three for the third, and so on. So for 20, it comes to $1 + 2 + 3 + \ldots + 20 = 210$ units of pay.

there remains $28\frac{4}{5}$ and the first had $28\frac{4}{5}$ so that we have found the first had $28\frac{4}{5}$ and the second had $38\frac{2}{5}$ and it is done.

36. There are two men that have denari. The first says to the second, "If I multiply mine by itself, I would have twice as much as you." The second says to the first, "If I multiply mine by itself, I would have three times as much as you." I want to know how much each had. Suppose that the first had a thing. And say that if he multiplies it by itself, that it would make twice that of the second. So multiplying a thing by itself makes a censo. So the second had as much as a censo and he [the first one] had twice as much as that of the second. So the second had of his $\frac{1}{2}$ censo. Now the second one said that if he multiplied his by itself, it would be 3 times that of the first. So multiply $\frac{1}{2}$ censo by itself, making $\frac{1}{4}$ censo di censo and this is 3 times as much as a thing, so it must be 3 things. Therefore you want to divide the things by the censi di censo, which is that you want to divide 3 by a $\frac{1}{4}$ which comes to 12. And we say that the thing is worth the cube root of 12. And we supposed that the first had a thing, so the first had the cube root of 12. And the second one we supposed to have $\frac{1}{2}$ censo, so know the value of the censo. Multiply the cube root of 12 by itself, making the cube root of 144. You have that the censo is worth the cube root of 144. Know the worth of the $\frac{1}{2}$ censo. You must divide the cube root of 144 by 2. Bring the 2 to the cube root, which is that you say 2 by 2 makes 4 and then 2 by 4 makes 8. Divide 144 by 8. It comes to 18. So the second one had the cube root of 18. And this way we have that the first had the cube root of 12 and the second had the cube root of 18 and it is done.

37. There were some men that had divided 60 denari among themselves, and each got as much at the other. Then two men joined and said, "We want our part and we agree that each gives 4 denari from their part." And this was the reasoning that the two who came second made to get what remained between those who first divided [the money]. I want to know how many men were there at first. Suppose that the first men were a thing. So you must divide 60 by a thing which comes to 60 divided by a thing and from what this comes to, each man gives 4. Now you

must know how much they gave the men since you know each gave 4. Multiply 4 by as many as were the first men, which is by a thing which we suppose it was, which makes 4 things. So you see that between all the first men, they gave 4 things, therefore take 4 things from 60, and remain with 60 less 4 things. Therefore 60 less 4 things remained with the first men and this is as much as they gave and they gave 4 things. Therefore 4 things are equal to 60 less 4 things. Restore the parts which is that you put onto each 4 things. You will have that 8 things are equal to 60. Divide 60 by 8. It comes to $7\frac{1}{2}$ and we say the thing is worth $7\frac{1}{2}$. And we suppose that the first men were a thing, therefore there were $7\frac{1}{2}$ men and it is done.

38. There were some men that had divided 60 florins among them and each had as much as the other. Now a man joined them and said that he wanted his part and they redivide these 60 florins from the beginning. And they found that those florins that came from the second [division] were such that, multiplying them by themselves, made as many denari as came from the first [division]. I want to know how many men were there at first. Do it this way: Suppose that the first men were a thing, so you must divide 60 by a thing which comes to 60 divided by a thing. Then join a man and divide again. So you want to divide 60 by a thing and one, which comes to a 60 divided by one thing and one. And say that multiplying this by itself makes as much as the event of 60 by a thing. So multiply 60 divided by a thing and one by itself, making 3600 divided by a censo and two things and one number and this is equal to 60 divided by a thing. And put this in order, how you see it drawn below, and then cross multiply, which is 60 by a censi and 2 things and one number, making 60 censi and 120 things and 60 numbers. Now multiply the other split, which is 3600 by a thing, making 3600 things and say that 3600 things are equal to 60 censi and 120 things and 60 numbers. Restore the parts, which is take from each part 120 things. So remain with 3480 things equal to 60 censi and 60 numbers. Divide by the censi, which is by 60. You will have that 58 things are equal to a censo and 1 number.

60	/	3600
by a thing		by a censo and two things and one number

Halve the things, which are 58, the half is 29, multiply it by itself, making 841, take away the number, which is one, remaining with 840 and we say that the thing is worth 29, which is the half of the things, plus the root of 840. And we supposed that the first men were a thing, therefore the first men were 29 plus the root of 840 and then there was one more, thus after there were 30 plus the root of 840 and do similar problems this way.

39. There are some men that have divided 60 florins among them and each got as much as the other. Now a man joined them and said that he wanted to divide this money over again with them. Now they came and they divided it over again and they found that multiplying that which they got at first against that of later made 20. I want to know how many men were there at first. Suppose that at first the men were a thing. You must divide 60 by a thing, which comes to 60 divided by a thing. Then say that one man came and they redivided from the start. So the second men were one thing and one man. So you must divide 60 by one thing and one, which comes to 60 divided by one thing and one. Now you want to multiply one against the other which is 60 divided by one thing by 60 divided by a thing and one, which is that you must multiply that which is over the line, one against the other, which is 60 by 60, making 3600. Now multiply that which is under the line, one against the other, which is one thing by one thing and one, which makes one censo and one thing. So done, multiplying one against the other, 3600 divided by one censo and one thing and this must make 20. Therefore multiply 20 by a censo and a thing, making 20 censi and 20 things equal to 3600. Divide by the censi, which are 20. You will have that one censo and one thing are equal to 180. Halve the things, which are one. It comes to $\frac{1}{2}$, multiplied by itself, making $\frac{1}{4}$. Join it to the number, which is 180. You have $180\frac{1}{4}$. And we say that the thing is worth the root of $180\frac{1}{4}$ less the half of the things which was $\frac{1}{2}$. And we supposed the first men were a thing, so the first men were the root of $180\frac{1}{4}$ less $\frac{1}{2}$. Then there was one more, so there were the root of $180\frac{1}{4}$ plus $\frac{1}{2}$

and it is done.

40. There are a few men that divided 60 florins among them so that one had as much as the other. And the they are joined by a man and he says that he wants his part and he said to redivide these 60 florins and redividing them they found that each got 6 florins less the second time. I want to know how many men were there at first. Suppose that the men there at first were a thing, so you want to divide 60 by a thing, which comes to 60 divided by a thing. And then say that you join a man, so that you want to divide 60 by a thing and one which comes to 60 divided by a thing and one. Therefore you must take 60 divided by a thing and one from 60 divided by a thing. And do it in this way: that you put 60 divided by a thing as a fraction and similarly put 60 divided by a thing plus one. Then cross multiply, which is 60 by a thing plus one, making 60 things and 60 numbers.

$$\frac{60}{\text{by a thing}} \quad / \quad \frac{60}{\text{by a thing and one}}$$

Now multiply across the other split, which is 60 by a thing making 60 things. Take 60 things from 60 things and from 60 numbers, remaining with 60 numbers. Now multiply that which is under the line against the other which is one thing by one thing and one. They make a censo and a thing. Then you must divide 60 by a censo and a thing, and it must come to 6. Therefore, multiply 6 by a censo and a thing making 6 censi and 6 things and therefore 6 censi and 6 things are equal to 60 numbers. You must divide by the censi, that is by 6, and you will have that one censo and one thing are equal to 10 numbers. Halve the thing, which is one. The half is $\frac{1}{2}$. Multiply it by itself making $\frac{1}{4}$. Join it to the number which is 10, making 10 and $\frac{1}{4}$ and you 10 and $\frac{1}{4}$. And we say that the value of the thing is the root of $10\frac{1}{4}$ less the half of the thing which was $\frac{1}{2}$. And we supposed that the number of men at first was a thing, so at first there were the root of $10\frac{1}{4}$ less $\frac{1}{2}$. After the men were one more, so there were the root of $10\frac{1}{4}$ plus $\frac{1}{2}$ and it is done.

41. There are two men that have denari. The first says to the second, "If you give me 12 of your denari, I would have twice as

Chapter 4: Comments on Additional Problems

In this chapter we offer comments and guides to problems that we found to be interesting because they are challenging to get hold of mathematically, or they show us the pedagogy of the time, or they give cultural insights, and, for one example, simply because it's fun.

The curricula of the abbacus schools concentrated on teaching the Hindu-Arabic number system, its arithmetic, and the application of that arithmetic to the increasingly complicated computational skills required by an evolving mercantile economy. Algebra was seldom addressed. However, treatises like our *Tratatto* reflected the pedagogy, if not the content, of the arithmetic schools. Phrases such as, "I want to say it like this," occur frequently and remind us that lessons were spoken. Students listened, perhaps took notes, and memorized. Clear and explicit instructions were given (and often repeated) for solving problems or using rules, but with no reasons as to why they worked. It was distinctly rote learning but with the expectation that students could readily use their skills in their professions. The first two examples of this chapter give us a window into the pedagogy of the time. The first gives instructions for dividing by "a number and the root of a number." We learn how to rationalize a denominator. Note the complicated example. The second introduces the addition of algebraic fractions. Note how the arithmetical example shows the way.

Example 1. Rule 13 of the rules for the arithmetic of roots starts:

> Now I want to show you how to divide by two roots,[1] which is to say, dividing a number by the root of number and a number. I want to say it like this: divide 100 by 10 and root of 20 and know what it comes to.

10 plus R of 20
X
10 less R of 20

[1]We think scribe erred and that the phrase should be "divide by two terms," not two roots.

> These are the two terms, which are 10 and the root of 20, that you must bring into one and bring them in this way, which is, you will multiply 10 and the root of 20 against its residue–that where it says "plus," you put "less"– which is 10 plus the root 20 by 10 less the root 20 and this "less" that you put, it is the residue...

This rule shows us how to divide a number by a binomial of the form $a + \sqrt{b}$. It demonstrates how with the following example: Divide 100 by $10 + \sqrt{20}$. The task is to find one number equal to $\frac{100}{10+\sqrt{2}}$. We would start by rationalizing the denominator. Why? So that the binomial in the denominator is replaced by a single number, in this case 80. We "bring it to one term." The instructions tell us how to construct the conjugate (here called "residue") of $10+\sqrt{20}$ by putting the word "less" where there was the word "plus" just as we replace the plus sign with a minus sign. The rest of the problem carries out in detail the steps needed to compute $\frac{100}{10+\sqrt{20}} \cdot \frac{10-\sqrt{20}}{10-\sqrt{20}}$. Rule 8 and its diagram are used to carry out the multiplication of the denominators, and Rule 1 is used for the numerators. The sequence of computations that follows will help you follow the rhetorical sequence in the text. We suggest that you start with the denominators and then move up to the numerators.

$$\frac{100}{10+\sqrt{20}} = \left(\frac{100}{10+\sqrt{20}} \times \frac{10-\sqrt{20}}{10-\sqrt{20}}\right) = \frac{100\times(10+\sqrt{20})}{100+10\sqrt{20}-10\sqrt{20}-20} = \frac{100\times(10+\sqrt{20})}{100+\sqrt{2000}-\sqrt{2000}-20} =$$
$$\frac{100\times(10+\sqrt{20})}{80} = \frac{1000}{80} - \sqrt{\frac{200{,}000}{80}} = 12\tfrac{1}{2} - \sqrt{\frac{200{,}000}{80}} = 12\tfrac{1}{2} - \sqrt{31\tfrac{1}{4}}$$

Example 2. Problem 13:

> Someone divided 100 by a quantity and then divided 100 by 5 more than the first and joining these two events together makes 20. I want to know by how much 100 is divided and by how much the next is divided.

The problem is simple in our terms: Solve for x: $\frac{100}{x} + \frac{100}{x+5} = 20$. But operations with algebraic fractions were rare at the time of our *Trattato*. Finding this example indicates that sources for our author were current and that algebra was slowly evolving. The instructions as to how to solve this problem also give us a look at the pedagogy and methodology of the time. The first task is to join (add) the fractions. He says:

> Now I want to show you something similar that is good advice for this joining and I say it this way: I want to join 24 divided by 4 with 24 divided by 6 that you see makes 10.

He then gives an example of adding two numerical fractions with the same numerator but different denominators that the reader would know the answer to. He gives detailed instructions that are transferable to the analogous problem with algebraic fractions.

> Therefore put 24 divided by 4 in the manner of a fraction which comes out to $\frac{24}{4}$. Again similarly, put 24 divided by 6 in the manner of a fraction. Then you must cross multiply, which is 24 by 6 making 144 and now multiply 4 by 24 which is over the 6. It makes 96, joined with 144 making 240. Now multiply what is under the line, which are 4 and 6, making 24. Then you must divide 240 by 24, which comes to 10.

This is good pedagogy no matter what the year. But the author had no alternative. Abstract variables were not yet in use. He could not demonstrate the general form by adding $\frac{a}{b} + \frac{a}{c}$. The constants stood in for the variables.

Using his model, he adds the algebraic fractions to obtain (in modern notation) : $\frac{200x+500}{x^2+5x} = 20$ and then multiplies through by the denominator, another rare operation. After restoring and balancing, he uses Rule 6 to find that the $x = \frac{5}{2} + \sqrt{31\frac{1}{4}}$.

Almost all of the problems in the *Trattato* that are not strictly arithmetical or algebraic, are about money—money lent, exchanged, shared, or swapped. Italy's mercantile economy was based on international trade supported by a vast shipping network. Each country and each city state of Italy had its own currency. Exchange was necessary. Investors needed capital. Borrowing and lending (with interest) was necessary, despite the church ban on usuary. Banking developed and would evolve into the complicated system it is today.

Example 3. The author presents this problem to illustrate the use of Rule 4. It is a typical lending problem. The goal is to find an interest rate.

> Someone lends another 100 lire for a term of 2 years, compounded at the end of the year, and when it comes to the end of two years 150 lire are returned to him. I ask at what rate per month was the lira lent. Suppose that the lira was lent at one thing of denari[2] per month. Therefore a lira makes 12 things of denari per year which is $\frac{1}{20}$ of a thing of lira. Therefore a lira makes $\frac{1}{20}$ things of lira per year so that 100 lire makes 100 by $\frac{1}{20}$ things of lire. And 5 things joined onto 100 lire makes 100 lire and 5 things and you have completed one year. And for the other year, multiply what comes at the end of the first year by the yield of the lira per year, that is, 100 plus 5 things by $\frac{1}{20}$ things which makes 5 things and $\frac{1}{4}$ censo which are what you want to join with 100 lire and 5 things, making 100 lire and 10 things and $\frac{1}{4}$ censo. And it is that much, between the interest and the capital, that 100 lire are [worth] in two years, which is 100 and 10 things and $\frac{1}{4}$ censo that are equal to 150 lire. Restore, that is remove 100 from each part, to get 10 things and $\frac{1}{4}$ censo equal to 50.
>
> Do it the way the rule says, that is to bring it to one censo, which is to divide by $\frac{1}{4}$, and you will have that one censo and 40 things are equal to 200 numbers. Halve the thing, it comes to 20. Multiply it by itself to make 400, join it to the number to make 600. Take the root which is a surd that presents as not a root.[3]
>
> Therefore we will say that the thing is worth the root of 600 less 20, which is the size of the thing. And we supposed that the lira was lent for a thing of denari per month. Therefore the lira was lent at the root of 600 denari less 20 denari per month and it is done.

The first complication of this problem is that there are two monetary

[2] 1 lira = 240 denari

[3] This means that 600 is not a perfect square.

denominations, the lira and the denaro, both used in Florence. For this problem, one lira is worth 240 denari. The second complication is that the interest is compounded once per year, but the problem asks for the rate that the money, 100 lire, is lent in denari per month per lire. The loan is compounded yearly over a period of two years. After two years 150 lire are returned. The author assumes that one lira earns a *thing* of denari each month. Over one year it earns 12 *things* of denari. Translating to lira, we find that one lira earns 12 divided by 240 lira per year, or $\frac{1}{20}$ *things* of lira per year. Since the original loan was 100 lire, it will earn $\frac{100}{20}$ or 5 *things* of lire the first year. At the end of the first year the loan is worth 100 lire and 5 *things* of lire. For the second year this total also earns $\frac{1}{20}$ *things* of lire per lire. Multiplying 100 and 5 *things* by $\frac{1}{20}$ of a *thing* yields 5 *things* and $\frac{1}{4}$ *censo*. Thus, at the end of the second year the loan is worth 100 lire and 10 *things* of lire and $\frac{1}{4}$ *censo* of lira. He knows that this is equal to 150 lire. So the author has the relation: $\frac{x^2}{4} + 10x + 100 = 150$. He restores and balances: $\frac{x^2}{4} + 10x = 50$. He solves with Rule 4. He finds that the rate is $\sqrt{600} - 20$ denari per lira per month, approximately 4.5 denari per month. This would come to 22.5% per year. Exorbitant?

A nuanced view of money and usury had emerged as well as financial strategies to dodge the accusation of usury. At all times, exorbitant interests rates were condemned. The difference as to whether lending with profit was usuary or not came to depend on its "use." It is a complicated picture but in short, if money were lent with interest to a pauper to buy food, then the use of that money cannot enable the pauper to pay back the principal much less the interest. Usuary. But if money is lent as capital for a profit-making venture, repayment with interest was tolerated.

Example 4. Problem 35 is an interesting problem solved without algebra. We are asked to find how many denari two men have given certain conditions. The author says he has a nice rule that will handle all such problems. He gives no indication of the source of this rule or its justification. What is remarkable is that algebraically, the problem is a set of three simultaneous equations in three variables but it is solved here with proportions only, no algebra. A big hint as to why it works can be found in Fibonacci's *Liber Abaci* in a problem from chapter 12, *A Problem on the Same Thing Proposed to Us by A Master near Constantinople.* He solves a problem similar to problem 35 using a line broken into four segments and applying his good knowledge of proportions. We will adapt his method to this problem in stages.

> [1] There are two men who have denari. The first says to the second, "If you give me 16 of your denari I will have twice as much as you." The second says to the first, "If you give me such a part of yours as I have given you of mine, I would have three times as much as you." I want to know how much each had. I want to teach you this problem by a nice rule by which you can do all similar problems. You know that [when] something is said to be twice as much, it is as two thirds of the whole quantity and three times as much as the said quantity is $\frac{3}{4}$ of the whole quantity. [2] Now know in what number one can find $\frac{2}{3}$ and $\frac{3}{4}$ which you find to be 12. Now take $\frac{2}{3}$ of 12 which is 8 and take $\frac{3}{4}$ of 12 which is 9, joined with 8, makes 17. Take away 12, in which you found the numbers, there remains 5 and 5 is our divisor. [3] Now multiply 12 in which you find the numbers, against what was asked for by the first from the second, which is 16, making 192, divided by 5, which you kept, it comes to $38\frac{2}{5}$ and 38 $\frac{2}{5}$ is what the second had. Now to know what the first had, do it this way. The first asked the second for 16 denari. Take 16 away from $38\frac{2}{5}$ remaining with $22\frac{2}{5}$. So the second remained with $22\frac{2}{5}$. And the first said he had twice as much as the second so it must be that he had $44\frac{4}{5}$. You know that the first had $44\frac{4}{5}$ with the 16 that the second gave him. Therefore take 16 from $44\frac{4}{5}$ and there remained $28\frac{4}{5}$ and the first had $28\frac{4}{5}$ so that we have found the first had $28\frac{4}{5}$ and the second had $38\frac{2}{5}$ and it is done.

[1] Draw a line segment *ae* with three intermediate points *b*, *c*, *d*.

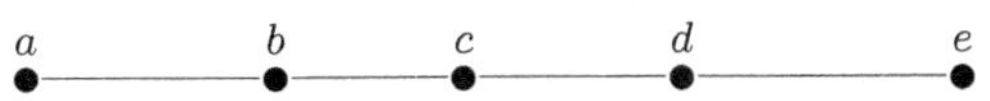

Let the segment *ac* be equal to the number of denari held by the first man, and *ce* be the number held by the second. The first segment,

ac, is cut at b. Let bc to be the number of denari that the first will give to the second, a number not given directly in the problem. Then ab is the number of denari left to the first man after he has given some denari to the second. The segment ce is cut at d. Then cd is the number of denari that the second man is to give to the first. So cd is 16. The length of only one of the four segments is known at the beginning. The goal is to find the lengths of the segments ac and ce, the number of denari held by the first man and the second man respectively.

The first part of the problem says: "If you give me 16 denari then I will have twice a much as you." The number of denari the first man would have after receiving the 16 denari is segment ad, the sum of ac and cd. Thus ad is twice de (the amount the second man has minus 16) and hence ad is $\frac{2}{3}$ of the whole segment ae. Now for the second part. We do not yet know how many denari the first is to give to the second but when he does, the second will have three times as many denari. Thus segment be is three times segment ab and hence be is $\frac{3}{4}$ of the whole segment ae.

[2] Now the author adds the two fractions, $\frac{2}{3}$ and $\frac{3}{4}$, to get $\frac{17}{12}$. What does this represent on the line? If we add the number of denari represented by the segments ad and be we get $\frac{2}{3}ae$ and $\frac{3}{4}ae$ or $\frac{17}{12}$ of the whole segment ae. The sum of ad and be is the same as the sum of ae and bd. Removing a copy of ae from each side gives us that the segment bd is $\frac{5}{12}$ of the whole segment ae. *Keep this in mind.* Note that ae represents the total number of denari held by the two men while bd represents the sum of the two amounts to be exchanged.

[3] At this point we need to look at the beginning of the problem again:

> The second says to the first, "If you give me such a part of yours as I have given you of mine."

The second man asks for an amount that is the same fractional part of the first man's money as 16 was of his own. In terms of our line, he asks that $bc{:}ac=cd{:}ce$ or rearranged, that $ac{:}ce = bc{:}cd$. *Keep this in mind also.*

The whole problem depends on a simple fact about proportions that Fibonacci and our author would have understood and probably used frequently: Suppose that we have the proportions, $A : B = C : D$ and we know that $k(A+B) = C+D$ for some positive number k. Then we have $kA = C$ and $kB = D$. Now we apply this fact to our problem. We know that $bd = \frac{5}{12}ae$. Since $bd = bc+cd$, and $ae = ac+ce$, we have $\frac{5}{12}(ac+ce) = bc+cd$. Since $ac : ce = bc : cd$ and $\frac{5}{12}(ac+ce) = bc+cd$, we have $\frac{5}{12}ac = bc$ and $\frac{5}{12}ce = cd$.

Since $\frac{5}{12}ce = cd$ and $cd = 16$, we have $\frac{5}{12}ce = 16$. We solve for ce, the number of denari the second man has, to get that $ce = \frac{192}{5} = 38\frac{2}{5}$. If the second man gives 16 denari to the first then the second will have $22\frac{2}{5}$ denari. According to the problem the first man will then have twice as much or $44\frac{4}{5}$ denari. Removing the 16 leaves the first man with $28\frac{4}{5}$ denari. The problem is done. The author solved the following system with no algebra:

$$\begin{aligned} x + 16 &= 2(y - 16) \\ y + z &= 3(x - z) \\ \tfrac{z}{x} &= \tfrac{16}{y} \end{aligned}$$

Example 5. When our *Trattato* was written, northern Italy was an aggregation of city states, each with its own government and currency. Problems 24 to 33 are about exchanging currencies. We find, for instance, florins (fiori d'oro or fiorini) from Florence exchanged for veneziani from Venice or tornese from Turin, or genovese from Genoa, all important commercial cities in a vast trading network. A vast network of banks developed to handle such exchanges. But even within a city, coins of different types were exchanged. Problem 25 is a good example. The exchange is between two coinages of Florence, the fiorini (florin) and the piccioli. Their relation was not as simple as quarters to dimes. Because the florin was gold and the piccioli was silver, their relative worth on a given day was based on the fluctuating value of gold vis a vis silver. Problem 25 asks for the profit (or fee) the money changer (tavoliere) made per florin exchanged.

> Problem 25. A merchant must have from another, who is a tavoliere [a money changer], 100 lire in fiorini and when it comes to the end of some period of time and this merchant returns to the tavoliere and says, "I want the exchange of

> those 100 lire in fiorini and write them as piccioli." And the tavoliere does so and gives him in the exchange for those 100 lire in fiorini and then he wrote that he must have 100 lire in piccioli. And when it came to the end of another period of time and this merchant returns to the tavoliere and says, "Those 100 lire that you wrote for me in piccioli, I want you to write them for me in fiorini." And the tavoliere does so and the merchant says how many lire in fiorini they are and he says, "They are as many as I gave you for the exchange of 100 lire in fiorini at first." I ask for the lira in fiorini that he received from the exchange of a lira of piccioli.
>
> Suppose that he [the tavoliere] receives a thing for an exchange, therefore the 1 lira of fiorini is worth 1 lira and 1 thing of piccioli. Now we know what the 100 lire of fiorini are worth, that they are worth 100 lire and 100 things of piccioli and the tovoliere wrote 100 lire of piccioli. So in the exchange what he made was 100 things and keep in mind 100 things.
>
> Now he says that he wants to know how much the 100 lire of piccioli that he wrote him are worth in fiorini. Therefore you must divide the 100 lire by 1 and 1 thing and I say that this must come to the 100 things that he got for the exchange. Then multiply one lire and one thing by 100 things, making 100 things and 100 censi and this is equal to 100 lire. Divide by the censi and you have that one thing and one censo are equal to 1 lire. Halve the thing, which is $\frac{1}{2}$, multiply it by itself, making $\frac{1}{4}$, and join it to the number. You will have $1\frac{1}{4}$ and you say that the the thing is worth the root of one and $\frac{1}{4}$ less $\frac{1}{2}$.
>
> And we supposed that, for the exchange of a lira in piccioli for a lira in fiorini, he gets one thing. Therefore he receives the root of one and $\frac{1}{4}$ less $\frac{1}{2}$ in fiorini from that piccioli which could be solde xii, denari vi and indeed anything else.

The 100 lire in florins is exchanged for 100 lire in piccioli. However there is an implicit fee: 1 florin is worth $1+x$ piccioli and so 100 fiorini are worth $100 + 100x$ in piccioli. But the tovioliere writes that the merchant has 100 lire in piccioli on deposit. The tavoliere's profit is thus 100 things. Upon the re-exchange, the merchant gets 100 piccioli which are $\frac{100}{1+x}$ fiorini. The merchant says that this number is the same as the tavoliere's fee, $100x$. And now we have the equation to solve: $\frac{100}{1+x} = 100x$.

The equation is a bit confusing to us. The units on the left and right are different with fiorini on the left and piccioli on the right. But the answer, the value of x, is $\sqrt{1\frac{1}{4}} - \frac{1}{2}$ or $\frac{\sqrt{5}-1}{2}$, the golden ratio, a value that would have been well known to the author of the problem. He would also know that if x is the golden ratio, then $x = \frac{1}{x+1}$. Then of course the $\frac{100}{x+1}$ (the value of the piccioli in fiorini) equals $100x$. (The latter just happens to be the profit or fee that the tavoiliere received). Perhaps the problem was formulated (rigged?) with the answer in mind.

About the word "tavoliere." The root of the word is tavola, a table or flat board. The money changer could be found at his table, perhaps covered with a green cloth. On it were stacks of coins and a book for keeping records. Ultimately, the book became the most important feature. Instead of carrying hard currency, always risky, the merchant would obtain a bill of exchange from the money changer ("write it for me" as our merchant says) to be cashed at the destination of the merchant. These tables evolved into lucrative banking centers based on bills of credit and exchange, agile and profitable. (Another old Italian word for a flat board is banca from which we get our word bank.)

Example 6. We close this chapter with a problem of a different sort from the others. Again there are unstated assumptions in the problem. But it leads us to discover that the author has used a formula very familiar to us. Here is problem 34.

> A well-digger must dig a well that is 20 braccia deep and he should get 20 lire. Now he came and the amount he dug got him xii [12] lire. I want to know how many braccia he dug. Reproducing the work braccio by braccio, do it this way: always collect one by one up to 20 which makes 210.

> Now say: From 210 dug, if I got 20 lire, how much was dug that got 12 lire? Multiply 12 by 210, making 2520, divide by 20. It comes to 126. And you say that he dug, digging one by one, to that number which makes 126. To know how much that number was, now collect one by one up to a thing, then you must do this: take $\frac{1}{2}$ of a thing, which is $\frac{1}{2}$ thing. Now put one onto the thing, making one thing and one. Now multiply $\frac{1}{2}$ thing by one thing and one, making $\frac{1}{2}$ thing and $\frac{1}{2}$ censo, and $\frac{1}{2}$ thing and $\frac{1}{2}$ censo is equal to 126. Divide by the censi, which is $\frac{1}{2}$. You will have that one censo and one thing is equal to 252. Halve the thing, which is one, the half of which is a $\frac{1}{2}$, multiply by itself, making $\frac{1}{4}$, join it to the number, making $252\frac{1}{4}$. And we supposed that the number was a thing therefore he dug the well to the root of $252\frac{1}{4}$ less $\frac{1}{2}$ braccio.

The word *braccio* (pl. *braccia*) means arm in Italian. As a unit of length, it is roughly 23 inches, the length of an arm. A well digger is to dig a well 20 braccia deep. For this work he should receive 20 lire. He does not complete the task. What he did dig earned him 12 lire. The key sentence is "Reproducing the work braccio by braccio, do it this way: always collect one by one up to 20 which makes 210." The assumption is that if the well digger earns a thing of lire for the first braccio dug, then he will earn 2 things of lire for the next braccio down, then three things, etc. Adding up the numbers from 1 to 20 gives 210. Thus he was to earn 210 things of lire if he dug the entire 20 braccia. The author recreates the formula for the sum of consecutive integers in terms of an unknown thing, namely that $1 + 2 + \ldots + x = \frac{x^2+x}{2}$. He knows that the digger has accomplished $\frac{12}{20}$ of the job and so he is paid $\frac{12}{20} \cdot 210 = 126$ lire. He then solves the equation we would write as $\frac{x^2+x}{2} = 126$. Using the rule appropriate to this situation he finds that the well digger has dug down $\sqrt{252\frac{1}{4}} - \frac{1}{2}$ or approximately 15.4 braccia. (29.5 ft).

Suggested Exercises

1. In Chapter 1, you will find a problem of Diophantus, namely "To divide a given square into two squares." We discuss the example he used to show his method of solution. The square he chose was 16. He then chose x^2 and $(2x - 4)^2$ as the two squares.

   (a) Why did he choose a term in which he subtracted 4?

   (b) Why didn't he choose the second to be $(2x + 4)^2$?

   (c) Could he have chosen the second to be $(x–4)$?

   (d) What would his solution have been if he had chosen x^2 and $(4x - 4)^2$?

   (e) Find a solution if the given square is 25.

2. There are three forms of quadratic equation treated in the manuscript. The rules for solution of these three forms are given in Rules 4, 5, and 6. In the language of the manuscript, they are:

   Rule 4. Censi and things equal to numbers
   Rule 5. Censi and numbers equal to things
   Rule 6. Censi equal to numbers and things

   The rules give algorithms for finding positive solutions. Two solutions are given by Rule 5.

   (a) Why is only one solution given by Rule 4?

   (b) Why is only one solution given by Rule 6? (Hint: the reasons are different.)

   (c) Are there any of these forms that, when applied, potentially have no solutions? If so, indicate which and why.

3. In most of the problems in our treatise an expression involving *censi*, *things*, and *numbers* is found to be equal to some other expression involving *censi*, *things*, and *numbers*. The two terms, *alcibra* and *amuchabile*, from the manuscript's title refer to two operations applied to these expressions: Subtracted terms are replaced by added terms by adding the appropriate quantities to each side and if two like kinds of quantities appear on each side of the equality, one is removed from the other. In this way

the problem is reduced to one solvable by one of the first six rules. In each of the following problems, reduce the expression to a standard form and determine which of the Rules 1 through 6 is appropriate to use to solve it.

(a) 5 censi and 3 things less 4 equals 3 censi less 3 things less 3

(b) 100 censi less 200 things plus 50 equals 200 censi and 200 things less 75.

(c) 2 censi and 5 equals one censo and 10.

(d) 5 censi equals one censi and 12 things.

4. In each of the following problems write a solution as the author of the treatise would have written it for his readers.

   (a) Make for me of 10 two parts such that adding the larger divided by the smaller to the smaller divided by the larger is 3.

   (b) Make for me of 10 two parts such that multiplying one by 5 makes as much as the other multiplied by the root of 2.

   (c) Make for me of 10 three parts in continued proportion such that multiplying the first against the second makes 3 and multiplying the second against the third makes 9. Note: x, y and z are in continued proportion if $x : y :: y : z$.

   (d) Make of 10 two parts such that the root of the larger times the root of the smaller is 3.

   (e) Make of 10 two parts such that the root of the larger plus the root of the smaller is 3.

5. In Problem 6 of the third section of the manuscript, the scribe makes an error that propagates through to the end of the problem. Fix the error and finish the problem correctly for him.

6. The following problems are from manuscripts written by other authors. The prefaces of these works along with some problems are found in the Appendix to this monograph. Each problem is quoted from the indicated manuscript, but without a solution. Try to find the solutions to the following:

(a) From Fibonacci's *Liber Abaci*, try the first five problems given. Not all require algebra. (We think you know the answer to the sixth problem. Do you recognize it?)

(b) From Jacopo da Firenze's *Tractatus Algorismii*, Problem 3.

(c) From Maestro Dardi's *Aliabraa Argibra*, problem 2. (This is a bit more challenging.)

(d) From Fillipo Calandri's *Trattato di Arithmetrica*, all four problems. None require algebra.

7. Problem 34 in the manuscript concerns the digging of a well.

> "A well-digger must dig a well that is 20 braccia deep and he should get 20 lire. Now he came and the amount he dug got him 12 lire. I want to know how many braccia he dug."

In problem 34 it is assumed that if the digger is paid a certain amount for the first braccio, he will be paid twice as much for the second braccio down, three times as much for the third down, and so on. Suppose instead that the sequence of costs per braccio are as the odd numbers. That is, he is paid an amount for the first braccio, three times that amount for the second braccio, five times that amount for the third braccio, and so on. Now how many braccia has he dug?

8. Problem 35 gives a rule for its solution but no justification for the rule. The rule does solve the problem (and all similar ones) correctly. The problem is:

> There are two men who have denari. The first says to the second, "If you give me 16 of your denari I will have twice as much as you." The second says to the first, "If you give me such a part of yours as I have given you of mine, I would have three times as much as you." I want to know how much each had.

There are three given quantities in the problem: 16 denari and multipliers of 2 and 3.

(a) Suppose that the problem read: "There are two men who have denari. The first says to the second, 'If you give me 20 denari then I will have 3 times as much as you.' The second says to the first 'If you give me such a part of yours as I gave you of mine, then I will have 6 times as much as you.' I want to know how much each had." Adopt the procedure given in problem 35 to solve this problem. Note: This problem is discussed in Chapter 4.

(b) Suppose that the problem read: "There are two men who have denari. The first says to the second, 'If you give me A of your denari I will have m times as much as you.' The second says to the first, 'If you give me such a part of yours as I have given you of mine, I would have n times as much as you.' I want to know how much each had." Assuming that A is a positive number and m and n are integers greater than 1, find how many denari the second man has in terms of A, m, and n.

9. Rule 12 gives an algorithm for solving "When the cubes are equal to things and a number". The rule is not always correct.

   (a) Find a numerical example for which the rule fails.

   (b) Find a numerical example for which the rule gives a correct solution. (Hint: Try an example with root 1.)

10. Find a solution for "A cube and 12 things equal 32." Use only the rule found in example 2 that follows the Cardano Preface to his *Ars Magna* found in the Appendix.

11. Cardano solves the cubic equation $x^3 + 6x = 20$ and finds the solution to be

$$\sqrt[3]{\sqrt{108} + 10} - \sqrt[3]{\sqrt{108} - 10}.$$

By inspection we see that the root is 2. How does one go from Cardano's expression to 2?

12. The problems that we find in pre-modern algebra texts, starting with those from Mesopotamia, have persisted as similar exercises for today's students. We find problems in which we divide

a number (or length) into two parts with certain constraints, problems involving interest rates, problem involving two towers or poles of different heights, or a task accomplished by several workers working at different speeds, and more. Your task is to go back to a pre-calculus or calculus text and find problems that echo the older ones. Solve them (again). You may need newer tools such as calculus not available to our authors.

13. We can see only part of Problem 41 on the last page of the *Trattato*. The problem is stated:

> There are two men that have denari. The first says to the second "If you give me 12 of your denari, I would have twice as much as you." The second says to the first, "If you give me such a part of yours that 7 of mine enter, I would have 3 times as much as you." I want to know how much he had.

The solution is started but not completed. Try to solve it without the guidance of our scribe. To start, let x be the amount that the first man has and y be what the second man has. We took the phrase "such a part of yours that 7 of mine enter" to mean the amount that the second man asks for is $\frac{x}{7y}$. With this interpretation, the answer is: $x = \frac{4}{35}\left(\sqrt{35221} - 124\right)$, $y = \frac{2}{35}\left(\sqrt{35221} + 191\right)$. Do you agree?

14. The website of the Mathematical Association of America (MAA) supports a section called Mathematical Treasures which are images and selected pages from mathematical manuscripts and texts from various libraries, museums, and private collections. You can find it at MAA Treasures:
https://web.archive.org/web/20240207092900/https://maa.org/press/periodicals/convergence/index-to-mathematical-treasures

Select a text. Write a brief essay on the topic of you selection and indicate on how your selection connects to abbacus tradition from which our *Trattato* comes. The text may have a direct connection (e.g. another work in the abbacus tradition) or it can be a parallel work from another culture (e.g. Hindu or Aztec).

References and Resources:

1. Al-Khwarizmi, *The Algebra of Mohammed ben Musa*, translated and edited by Frederic Rosen, London: Nachdruck der Ausgabe, 1831. (Reprinted 1997 as IM&A volume 1.)

2. Anonimo, 1994. *Trattato dell'alcibra amuchabile: Dal Codice Ricc. 2263 della Biblioteca Riccardiana di Firenze*, ed. Annalisa Simi. Siena: Servizio Editoriale dell'Universita di Siena (Quaderno 22).

3. Cardano, Gerolamo, *Rules of Algebra, 1545* (*Artis Magnae, sive de regulis algebraicis*), Trans. Richard Widmer, Dover 2007

4. Epstein, Steven, *An Economic and Social history of Later Medieval Europe, 1000-1500*, Cambridge University Press, 2009.

5. Fibonacci, *Fibonacci's Liber Abaci: A translation into modern English of Leonardo Pisano's Book of Calculation*, 2002, tr. L. E. Sigler. New York: Springer.

6. Francci, Raffaella. "The History of Algebra in Italy in the 14th and 15th Centuries. Some Remarks on Recent Historiography" Actes D'Historia de la Ciencia i de la Tecnica, vol. 3, no. 2, 2010, pp. 175–194.

7. Grendler, Paul F. *Schooling in Renaissance Italy: Literacy and Learning, 1300–1600.* Baltimore Johns Hopkins University Press. 1989.

8. Heath, Sir Thomas L. *Diophantus of Alexandria: A Study in the History of Greek Algebra.* 2nd ed., Cambridge University Press, 1910

9. Høyrup, Jens. *Lengths, Widths, Surfaces: A Portrait of Old Babylonian Algebra and Its Kin.* Springer-Verlag, 2002

10. Hoyrup, Jens, *Jacopo da Firenze's Tractatus Algorismi and Early Italian Abbacus Culture.* Basel: Birkhauser, 2007

11. Hoyrup, Jens. "Hesitating Progress – the Slow Development toward Algebraic Symbolization in Abbacus- and Related Manuscripts,

C.1300 to C.1550." *Max Planck Instititute for the History of Science*, 2009, Pre-print 390

12. Høyrup, Jens. *The World of the Abbaco.* Birkhäuser, 2024. SpringerLink, https://doi.org/10.1007/978-3-031-25164-1.

13. Katz, V*ictor J. A History of Mathematics.* 3rd ed., Pearson, 2009.

14. Oaks, Jeffrey A. "Polynomials and Equations in Medieval Italian Algebra." *Bollettino di Storia delle Scienze Matematiche*, vol. 30, 2010, 23-60.

15. Petrocchi, A. (2016). "Mercantile arithmetic in Renaissance Italy: A translation and study of selected passages from a vernacular Abbaco work." *Ceræ*, 3.

16. Witt, Ronald G. "What Did Giovannino Read and Write? Literacy in Early Renaissance Florence." I Tatti Studies in the Italian Renaissance, vol. 6, 1995, pp. 83-114

The final two resources do not come to us via mathematics. Though quite different, they both offer extraordinary insight into the times and values of the 14th and 15th centuries in Italy. The first, *The Merchant of Prato*, by Iris Origo, delves into the life of the merchant Francesco Datini from his first venture into the merchantile life at age 15 to his death at age 75. Datini, compulsive and exact, left a vast collection of records and letters. These were distilled by Origo into chapters that tell us of his life and his businesses though which we see the broader terrain of mercantile life. We also glimpse the horror of the plague. The Black Death arrived in Italy in 1347 and recurred regularly. It did not shut down commerce.

The second offers us a visual tour. The website presents Ambroglio Lorenzetti's masterpiece, *Allegoria ed effetti del Buono e del Cattivo Governo*, The Allegory and Effects of Good and of Bad Government. Painted in 1345, his frescoes cover the walls of the Sale de Nove, Room of Nine, in which the city council of Siena met. On one of the side walls we see both city and country life depicted in times of peace and prosperity, the effects of good government. On the other, the dire effects of bad government. The room is in the Palazzo Publico, or Town Hall of Siena.

17. Origo, Iris, *The Merchant of Prato: Francesco di Marco Datini,* 1335-1410. Alfred A. Knopf, 1957.

18. A. Lorenzetti's *Good and Bad Government,*: https://www.wga.hu/html_m/l/lorenzet/ambrogio/governme/index.html

Appendix: Prefaces to Six Pre-Algebra Works

This appendix contains the prefaces of six works related to the algebra of the abbacus tradition. In them, the authors introduce their works. However you will not find mathematics in these introductions. Rather they give a glimpse of the cultural context in which the works were written. In them we find prayers, invocations, philosophical statements, and dedications. And we find enthusiasm. The works bracket the era of the abbacus school algebra. We start well before the schools existed in Italy with the seminal work of al-Khwarizmi, his *Al-Jabr*, written in approximately 800 CE. It is the source of the algebra found in Fibonacci's *Liber Abaci*, a work that helped bring Arab and Hindu arithmetic and algebra to Italy. Fibonacci's preface tells of a personal journey. Our manuscript has no preface or introduction so instead we present the introductions to two manuscripts from Italy that were roughly contemporary with the *Trattato*. They are from Jacopo da Firenze's *Tractatus Algorismii*, c. 1307, and from Master Dardi's *Aliabraa Argibra*, c. 1344. They are followed by the introduction to a work written at the end of the 15th century, Fillipo Calandri's *Arithmetica*, which was mass produced on the newly invented printing press. Finally, we end with the preface to Cardano's *Ars Magna*, the book that marks a leap forward into the modern era of algebraic thought. Upon reading these excerpts, you can see how much changed over 400 years but also how much stayed constant.

Al-Jabr by al-Khwarizmi

Around 820 CE, al-Khwarizmi wrote his *al-Kitāb al-Mukhtaṣar fī isāb al-Jabr wal-Muqābalah* (**The Concise Book of Calculation by Restoration and Balancing**) or more simply, the *Al-Jabr.* The translation by Friedrich Rosen from 1831 remains a principal English translation of the *Al-Jabr.* What follows is an excerpt from Rosen's preface to his translation entitled *The Algebra of Mohammed Ben Musa.*(See Reference 1.) In it, he sets the context for al-Khwarizmi's work. Rosen's preface is followed by the preface written by al-Khwarizmi.

From Rosen's Preface:

> In the study of history, the attention of the observer is drawn by a peculiar charm towards those epochs, at which nations, after having secured their independence externally, strive to obtain an inward guarantee for their power by acquiring eminence as great in science and in every art of peace as they have already attained in the field of war. Such an epoch was, in the history of the Arabs, that of the Caliphs al-Mansur, Harun al-Rachid, and al-Mamun, the illustrious contemporaries of Charlemagne; to the glory of which era, in the volume now offered to the public, a new monument is endeavoured to be raised.

Two pages from a 14th century copy of al-Khwārizmī's Algebra, Oxford, Bodleian Library MS. Huntington 214

Al-Khwarizmi's Preface, as translated by Rosen:

> In the Name of God, Gracious and Merciful! This book has been composed by Mohammad ibn Musa al-Khwarizmi. He commences it thus: Praised be God for his bounty towards those who deserve it by their virtuous acts: in performing which, as by him prescribed to his adoring creatures, we express our thanks, and render ourselves worthy of the continuance (of his mercy), and preserve ourselves from change: acknowledging his might, bending before his power, and revering his greatness! He sent Muhammad (on whom may the blessing of God repose!) with the mission of a prophet, long after any messenger from above had appeared, when justice had fallen into neglect, and when the true way of life was sought for in vain. Through him he cured of blindness, and saved through him from perdition, and increased through him what before was small, and collected through him what before was scattered. Praised be God our Lord! and may his glory increase, and may all his names be hallowed besides whom there is no God; and may his benediction rest on Muhammad the Prophet and on his descendants! The learned in times which have passed away, and among nations which have ceased to exist, were constantly employed in writing books on the several departments of science and on the various branches of knowledge, bearing in mind those that were to come after them, and hoping for a reward proportionate to their ability, and trusting that their endeavours would meet with acknowledgment, attention, and remembrance, content as they were even with a small degree of praise; small, if compared with the pains which they had undergone, and the difficulties which they had encountered in revealing the secrets and obscurities of science. Some applied themselves to obtain information which was not known before them, and left it to posterity; others commented upon the difficulties in the works left by their predecessors, and defined the best method (of study), or rendered the access (to science) easier or placed it more within reach; others again discovered mistakes in preceding works, and arranged that

> which was confused, or adjusted what was irregular, and corrected the faults of their fellow labourers, without arrogance towards them, or taking pride in what they did themselves. That fondness for science, by which God has distinguished the Imam al- Mamun, the Commander of the Faithful (besides the caliphat which He has vouchsafed unto him by lawful succession, in the robe of which He has invested him, and with the honors of which He has adorned him), that affability and condescension which he shows to the learned, that promptitude with which he protects and supports them in the elucidation of obscurities and in the removal of difficulties, has encouraged me to compose a short work on Calculating by (the rules of) Completion and Reduction, confining it to what is easiest and most useful in arithmetic, such as men constantly require in cases of inheritance, legacies, partition, law-suits, and trade, and in all their dealings with one another, or where the measuring of lands, the digging of canals, geometrical computation, and other objects of various sorts and kinds are concerned-relying on the goodness of my intention therein, and hoping that the learned will reward it, by obtaining (for me) through their prayers the excellence of the Divine mercy, in requital of which, may the choicest blessings and the abundant bounty of God be theirs. My confidence rests with God, in this as in everything, and in Him I put my trust. He is the Lord of the Sublime Throne. May His blessing descend upon all the prophets and heavenly messengers!

After his preface, al-Khwarizmi begins his compendium, "Completion and Reduction," with the following observation:

> "*When I considered what people generally want in calculating, I found that it always is a number.*"

Al-Khwarizmi presents six rules for solving linear and quadratic equations. They are listed in Chapter 1, on page 2. They will remain the core of algebra until the sixteenth century. He then gives six numeric examples, two of which are below. These are followed by a large set of numerical and applied problems. Most of his applied problems deal with property and legacies. They often assume some knowledge of Islamic law. Two are stated below.

Problems:

1. I have divided ten into two portions; I have multiplied the one of the two portions by the other; after this I have multiplied the one of the two by itself, and the product of the multiplication by itself is four times as much as that of one of the portions by the other. (Goal: Find the portions.)

2. I have divided ten into two parts; I have then multiplied each of them by itself, and when I had added the products together, the sum was fifty-eight dirhems.[1] (Goal: Find the parts.)

3. ON LEGACIES: On Capital, and Money Lent. A man dies, leaving two sons behind him, and bequeathing one-third of his capital to a stranger. He leaves ten dirhems of property and a claim of ten dirhems upon one of the sons. (His answer: This gives five dirhems, which is the thing that is taken out of the debts.)

4. Suppose that a man on his death-bed emancipates a slave, whose price is three hundred dirhems, not having any property besides; then the slave dies, leaving three hundred dirhems and a daughter. The problem is to find the legacy to the slave.

The translator Friedrich Rosen presents us with interesting additional information such as the following:

> Mohammad (AK) gives no definition of the science which he intends to treat of, nor does he explain the words *jebr*, and *mokabalah*, by which he designates certain operations peculiar to the solution of equations, and which, combined, he repeatedly employs as an expression for this entire branch of mathematics. ... I shall here subjoin a few remarks on its original sense, and on its use in Arabic

[1]The word dirhems (or now dirhams) was and is a unit of money in North Africa. In the arithmetic and algebra, it denoted a numerical value. So the problem amounts to $x^2 + (10 - x)^2 = 54$.

mathematical works. The verb *jabar* of which the substantive *jebr* is derived, properly signifies to restore something broken, especially to cure a fractured bone .... Hence the Spanish and Portuguese expression *algebrista* for a person who heals fractures, or sets right a dislocated limb.[2]

Reference:

1. Al-Khwarizmı, *The Algebra of Mohammed ben Musa*, translated and edited by Frederic Rosen, London: Nachdruck der Ausgabe, 1831. (Reprinted 1997 as IM&A volume 1.)

[2]From Google AI: Old Spanish, a person who heals fractures was called an algebrista. This term was used during the Renaissance and later to describe folk practitioners specializing in treating fractures, dislocations, and other musculoskeletal conditions using manipulation.

Fibonacci's *Liber Abaci* or Leonardo Pisano's *Book of Calculations*

Fibonacci wrote his *Liber Abaci* in 1202 and revised it in 1228. The preface presented here is from the 1228 revision. The book itself is a remarkable feat. He starts with an introduction to Indo-Arabic numerals, not yet used in Italy, and then he provides instruction in the placement notation for numbers and in the arithmetic it entails. He presents a wealth of arithmetic techniques with applications pertinent to the vibrant mercantile culture he lived in. The book is large. The English translation runs to over 600 pages. Of its fifteen chapters, only the last addresses algebra. The *Liber Abaci* was first translated into English in 2002 by Laurence Sigler, Professor of Mathematics at Bucknell University. It is the only English translation of this important and influential work. We use his translation. (See Reference 5.) Here is an excerpt from Sigler's introduction:

> *Liber abaci* is one of the most important books on mathematics of the Middle Ages. Its effect was enormous in disseminating the Hindu number system and the methods of algebra throughout Europe. This is the first translation of the Latin manuscript of *Liber Abaci* into a modern language. It is hoped that its availability to historians, mathematicians, and the public in general will make a contribution to their knowledge of this part of our cultural heritage. Mathematics and science are, after all, as much a part of our culture as literature, art, and music. It is as important for a person to know about the classics of mathematics and science as it is to know about the classics of literature and art.

Fibonacci's Dedication and Preface:

> You, my Master Michael Scott,[1] most great philosopher, wrote to my Lord[2] about the book on numbers which some time ago I composed and transcribed to you; whence complying with your criticism, your more subtle examining circumspection, to the honor of you and many others I with advantage corrected this

[1]Michael Scott was a philosopher and scholar with scientific interests in the court of Frederick II. He is mentioned in uncomplimentary terms in Dante's *Inferno*. There is a park near the river Arno in Pisa named for Michael Scott.

[2]The lord referred to here is Frederick II, the Emperor of the Holy Roman Empire in the 13th century. He was a patron of mathematics, science, and the arts.

work. In this rectification I added certain necessities, and I deleted certain superfluities. In it I presented a full instruction on numbers close to the method of the Indians, whose outstanding method I chose for this science. And because arithmetic science and geometric science are connected, and support one another, the full knowledge of numbers cannot be presented without encountering some geometry, or without seeing that operating in this way on numbers is close to geometry; the method is full of many proofs and demonstrations which are made with geometric figures. And truly in another book that I composed on the practice of geometry I explained this and many other things pertinent to geometry, each subject to appropriate proof. To be sure, this book looks more to theory than to practice. Hence, whoever would wish to know well the practice of this science ought eagerly to busy himself with continuous use and enduring exercise in practice, for science by practice turns into habit; memory and even perception correlate with the hands and figures, which as an impulse and breath in one and the same instant, almost the same, go naturally together for all; and thus will be made a student of habit; following by degrees he will be able easily to attain this to perfection. And to reveal more easily the theory I separated this book into xv chapters, as whoever will wish to read this book can easily discover. Further, if in this work is found insufficiency or defect, I submit it to your correction.

As my father was a public official away from our homeland in the Bugia customs house established for the Pisan merchants who frequently gathered there, he had me in my youth brought to him, looking to find for me a useful and comfortable future; there he wanted me to be in the study of mathematics and to be taught for some days. There from a marvelous instruction in the art of the nine Indian figures, the introduction and knowledge of the art pleased me so much above all else, and I learnt from them, whoever was learned in it, from nearby Egypt, Syria, Greece, Sicily and Provence, and their various methods, to which locations of business I travelled considerably afterwards for much study, and I learnt from the assembled disputations.

But this, on the whole, the algorithm and even the Pythagorean arcs, I still reckoned almost an errror compared to the Indian method. Therefore strictly embracing the Indian method, and attentive to the study of it, from mine own sense adding some and some more still from the subtle Euclidean geometric art, applying the sum that I was able to perceive to this book, I worked to put it together in xv distinct chapters, showing certain proof for almost everything that I put in, so that further, this method perfected above the rest, this science is instructed the eager, and to the Italian people above all others, who up to now are found without a minimum. If, by chance, something less or more proper or necessary I omitted, your indulgence for me is entreated, as there is no one who is with fault, and in all things is altogether circumspect.

Problems from Fibonacci's Liber Abaci. The first two are from chapter 15 and are done with algebra. The remainder are done with no algebra.

1. If you wish to separate 10 into two parts so that their product makes one fourth of the greater part multiplied by itself, then you put for the greater part the root that you will call the thing; there will remain for the smaller part 10 minus the thing, and it, multiplied by the thing, yields 10 things minus the census; and the multiplication of the thing by itself yields the census, because when the root is multiplied by itself the square of the root results; therefore ten things minus the census are equal to a fourth part of the census. [The problem is to find the parts.]

2. I divided 60 by a number of men, and each had an amount, and I added two more men, and I divided the 60 by all of them, and there resulted for each $2\frac{1}{2}$ denari less than that which resulted first. [The problem is to find the number of men in the first group.]

3. Two birds were above the height of two towers; one tower was 40 paces in height and the other 30, and they were 50 paces apart; at an instant the pair of birds descended flying to[wards] the center where there was a fountain, and they arrived at the same moment at the fountain which was between both towers. From the moment they left until the moment they arrived they flew in straight lines from the tops of the towers to the center of the fountain; the flights were of equal lengths. [The problem is to find how far the fountain is from each tower.]

4. A certain man buys 4 pieces of cloth for 80 bezants.[3] He buys the first for a certain price, and he buys another for $\frac{2}{3}$ the price of the first. He truly buys the third for $\frac{3}{4}$ the price of the second. Moreover the fourth he buys for $\frac{4}{5}$ the price of the third. It is sought how much each piece is worth.

5. A certain lion eats one sheep in IIII [4] hours, and a leopard eats one sheep in 5 hours, and a bear eats one sheep in 6 hours; it is sought, if one sheep is thrown to them, how many hours it will take them together to devour it?

6. **(The famous rabbit problem!)**A certain man had one pair of rabbits together in a certain enclosed place, and one wishes to know how many are created from the pair in one year when it is the nature of them in a single month to bear another pair, and in the second month those born to bear also. Because the abovewritten pair in the first month bore, you will double it; there will be two pairs in one month. One of these, namely the first, bears in the second month, and thus there are in the second month 3 pairs; of these in one month two are pregnant, and in the third month 2 pairs of rabbits are born, and thus there are 5 pairs in the month; in this month 3 pairs are pregnant, and in the fourth month there are 8 pairs, of which 5 pairs bear another 5 pairs; these are added to the 8 pairs making 13 pairs in the fifth month; these 5 pairs that are born in this month do not mate in this month, but another 8 pairs are pregnant, and thus there are in the sixth month 21 pairs; to these are added the 13 pairs that are born in the seventh month; there will be 34 pairs in this month; to this are added the 21 pairs that are born in the eighth month; there will be 55 pairs in this month; to these are added the 34 pairs that are born in the ninth month; there will be 89 pairs in this month; to these are added again the 55 pairs that are born in the tenth month; there will be 144 pairs in this month; to these are added again the 89 pairs that are born in the eleventh month; there will be 233 pairs in this month To these are still added the 144 pairs that are born in the last month; there will be 377 pairs, and this many pairs are produced from the above written pair in the mentioned place at the end of the one year.

The image below is a facsimile of the page from the *Liber Abaci* that presents the rabbit problem. The right column contains the first few Fibonnaci numbers. No more are needed for the given problem. Note that it

[3]A bizant is a coin from the middle east originally minted in Byzantium.

starts a bit differently from where we start the sequence today. Fibonacci did not explore his now famous sequence any further.

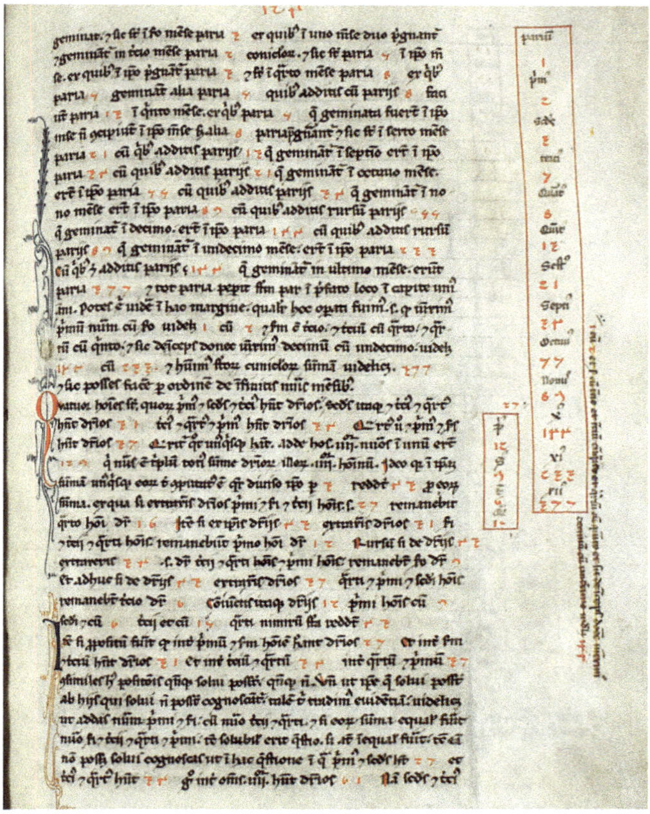

Fibonacci's Rabbit Problem
From a facsimile housed in the Museo Galileo, Biblioteca Nationale Centrale de Firenze, BNCF,
Conv. Soppr. C. 1.2616

Incipit and Preface to Jacopo Da Firenze's *Tractatus Algorismii*

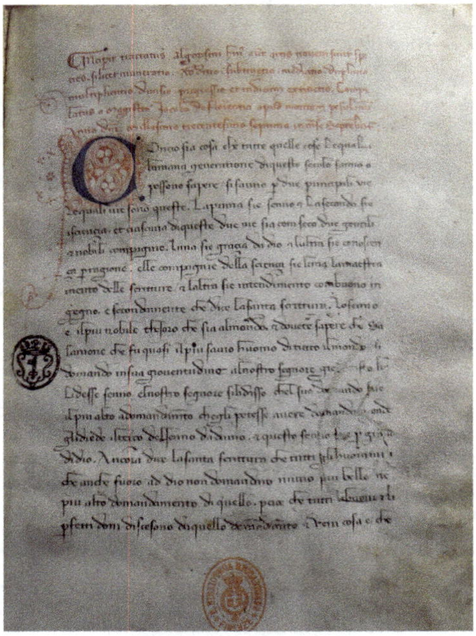

The first page of Jacopo's *Trattatus*.
Biblioteca Riccardiana, Florence, Ricc. 2236,
folio by permission of the Ministry of Culture

We let Jacopo introduce himself with his own Incipit.[1] The translation is from Hoyrup [2].

Incipit: The treatise of algorism begins, which art consists of nine species, namely, numeration, addition, subtraction, (mediation), duplation, multiplication, division, progression, and root extraction. Composed by master Jacopo of Florence in Montpellier, in the year 1307 in the month of September.

[1] *Incipit* is the Latin word for begin. It is used to denote the opening words of a text. In the Jacopo manuscript, the incipit is in Latin and the rest is in Italian.

Preface: Admittedly, all those things which the human race of this world know and are able to know, are obtained in two main ways, which ways are these. The first is discernment, the second is science.[2] And each of these two ways is accompanied by two gentle and noble partners. One is the grace of God. And the other is knowledge by reason. And of the partners of science, one is mastery of what has been written. And the other is understanding with good intelligence. And according to what the Holy Scripture says, discernment is the noblest treasure that there is in the world. And you shall know that Solomon, who was close to being the wisest man of all the world, asked the Lord in his youth to give him discernment. And our Lord said to him that his request was the highest request that he could have asked. Wherefore he gave him one third of the discernment of Adam, and this discernment was by grace of God. The Holy Scripture also says that no man until now asked God for any request more beautiful or higher than that, since all God's good and pure gifts descend from this request. It is true that one may call discernment and science, one natural discernment, the other accidental[3] science. And you shall know that everything men do naturally and by accident, our Father has granted (them) to know in his most holy virtue and grace and compassion. And therefore we are all obliged to thank Him who is such a sweet Father and Lord, who has given us to know so much subtlety. And therefore in His most holy name and His most holy honor we begin our treatise, which is called algorism. And know that it is called algorism because this science was first made in Arabia, and those who found it were similarly Arabs. And art in Arabic is called algo, and the number is called rismus, and so it is called algorism.[4] Which falls into five chapters, which I shall show you manifestly in our treatise ordered according to the said matter, as the said science asks for. And we begin in the honor and reverence of our Lord Jesus Christ and his most holy mother Virgin Mary and the whole celestial court, and with the assistance of our predecessors, and in honor of all masters and scholars of this science. And of every other honest person who might see and read this treatise with dedication and sense.

---

[2]Discernment and science: In the manuscript, the word that is here translated as discernment is "senno." It can also be translated as wisdom. Science is translated from the word "scienza," a word that, in Jacopo's time, denoted knowledge that is obtained through systematic study and deduction.

[3]Accidents: Medieval philosophers and theologians distinguished between the essence of a substance—that which it is —and accidents, those qualities that were nonessential to its identity.

[4]Note the questionable etymology.

The precedence argument

In Jacopo, we may or may not find the first abbacus manuscript that addressed algebra. Only about a third of the abbacus manuscripts had an algebra section and far fewer were exclusively dedicated to algebra. In Jacopo's beautiful manuscript, we can find a wealth of arithmetic techniques. As typical of the genre, these included the rule of three, and false position, and progression with examples that, like our *Trattato* and almost all others, send us into the contemporary mercantile world. The algebra section appears in only one of the three complete copies of the manuscript known to exist today, one in each of Florence, Milan and the Vatican. The first one can be dated from the beginning of the 14th century, the second and third from the beginning and the mid of the 15th century, respectively. The algebra section appears only in the Vatican manuscript. Historians disagree as to whether or not the algebra section was in Jacopo's original text or added later by a scribe and written by another author. Knowing one way or the other informs the thinking on the timeline of the evolution of algebra in the west: where and by whom did algebra first make its appearance in the west? The origin of algebra in Europe is itself complicated. For instance, just how much is owed to reworkings of Fibonacci's *Liber Abaci*? Clearly there were other sources, but who, what, and where from? That Jacopo wrote from Montpellier suggests possible influence from France and Spain. The argument that the first appearance of algebra was in Jacopo's 1307 treatise is advocated by Jens Hoyrup [2]. But we find both Raffaella Franci [1] and Jeffrey Oaks [3] argue that the algebra section was inserted later and that a better candidate for the first appearance of algebra was in Paolo Gheradi's *Liber di ragioni* from 1328, also written in Montpellier. Others get caught in the fray. Annalisa Simi, the transcriber of our *Trattato,* asserts that the problems associated with rules 1 through 6 are original and not found in the earlier text by Gheradi. Yet they are found almost verbatim in the algebra section of the Vatican manuscript of Jacobo's *Tractatus.* (See problems 1 and 2 below.) If indeed, the algebra in Jacopo was written in 1307, it would antedate our *Trattato.*

Problems from Jacopo's *Tractatus*:

1. Example to the said rule. [The "said rule" is Jacopo's Rule 5 which is Rule 5 in our *Trattato* as well.] And I want to say thus, make two parts of 10 for me, so that when the larger is multiplied against the smaller, it shall make 20. I ask how much each part will be.

2. (Also an example of Rule 5 in the *Trattato.*) Somebody makes two voyages, and in the first voyage he gains 12. And in the second voyage he gains at that same rate as he did in the first. And when he had completed his voyages he found himself with 54, gains and capital together. I want to know with how much he set out.

3. Somebody stays in a warehouse[5] 4 years. And in the first year and the fourth together he got 90 gold fiorini. And in the second year and the third together he got 60 gold fiorini. I want to know what resulted for him, each one by itself. And let them be in proportion and let the first be such part of the second as the second of the third, and as the third of the fourth.

4. I have ounces 60 of gold which contains carats 16 per ounce, and I want put it in fire and refine it so much that it becomes of carats 21 per ounce. I want to know how much these 60 ounces will become in weight, taken out of the fire when it is of carats 21, as I have said to you, neither more nor less.

References:

1. Franci, R., "The History of Algebra in Italy in the 14th and 15th Centuries. Some Remarks on Recent Historiography." *Actes D'Historia de la Ciencia I Nove Epoca/* Volume 3/2010, p. 175-194

2. Hoyrup, J. *Jacopo da Firenze's Tractatus Algorismi and Early Italian Abbacus Culture,*(Basel: Birkhauser, 2007)

3. Oaks, J. A. (2009), "Essay Review: Medieval Italian. Practical Mathematics", CHSPM/SCHPM Bulletin. November 2009.

[5]From the twelfth century onwards the maritime powers took advantage of the Arab system of funduqs, fondachi in Italian, developed for the caravan trade. They are warehouses that were not only for storing goods, but providing safe havens and accommodations for the merchants en route. They became centers of commerce. Jacopo has a number of fondachi problems like the one above. [4]

4. Peter Spufford, 2014. "The Provision of Stable Moneys by Florence and Venice, and North Italian Financial Innovations in the Renaissance Period," Financial and Monetary Policy Studies, in: Peter Bernholz Roland Vaubel (ed.), Explaining Monetary and Financial Innovation, edition 127, pages 227-251, Springer.

Manuscript locations:

1. Ricc. 2236, Biblioteca Riccardiana (Firenze);
2. Ms 90, Biblioteca Trivulziana (Milano);
3. Vat. Lat. 4826, Biblioteca Apostolica Vaticana.

Aliabraa Argibra by Maestro Dardi da Pisa

The *Aliabraa Argibra*, written by Maestro Dardi da Pisa around 1344, is the first known work in Italian that addresses algebra only. It is the work of an excellent mathematical mind, insightful, accurate and thorough. Dardi gives 194 numbered rules for solving algebraic equations rather than 24 listed in our *Trattato*. His treatment of a cubic equation that is not reducible to a quadratic equation, while not giving a general rule, presents a set of steps that gives the correct answer to his stated problem. Here is his preface:

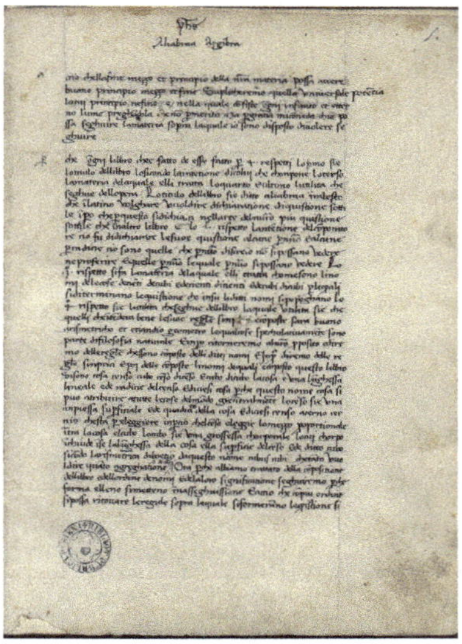

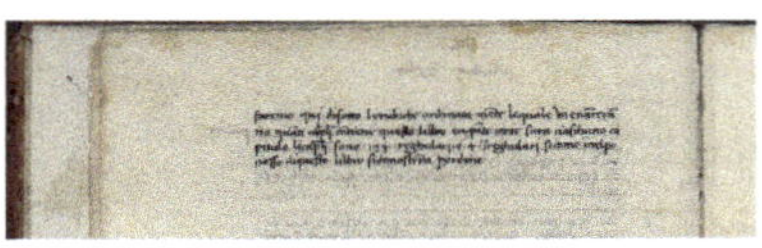

Maestro Dardi's Preface

YHS[1]

Aliabraa Argibra

So that the end, middle and beginning of our subject may have a good beginning, middle and end, we will beseech that universal power whose beginning is its end and in which consists every infinite and eternal light, praying that it grant me, not through merit but through grace, that I may continue the subject upon which I am disposed to want to continue. It is the case that every book that is made must be made with four attributes. The first is the title of the book, the second is the intention of the one who writes it, the third is the matter which it treats, the fourth and last is the usefulness that comes from the work. The title of the book is called aliabraa in Arabic which in vernacular Latin means the statement of subtle problems because by this, more subtle problems are stated in the art of numbers than in any other book. And for the second attribute, the intention of the author, which was to state his problems, some about number and some about roots, which are those which cannot be understood or explained by discrete numbers and those that by number, one can understand by number. The third attribute is about the subject that he treats, which are the names of things, of cenzi, of cubes, and of cenzi di cenzi, and cubi di cubi, by which one solves the problems that were proposed with the said names. The fourth attribute is the usefulness that comes from the book, which usefulness is that he who will understand well its simple and composite rules will be a good arithmetician and also a geometer for whom speculative things are part of natural philosophy. (Author's translation from from Franci, Quarderno 26. See reference 2 below.)

Epilogue: What's in a Name?

Before the 1980s there were three known copies of the *Aliabraa Argibra* in Italy, one in each of Siena, Florence, and Rome. None of the three provides an author's name or a date. In 1984, Warren Von Egmond [4] suggested that the author and date could be identified via a manuscript now housed in Paris (See Manuscript Location 1). The

[1]YHS, or more commonly, IHS is a Christogram, i.e. a set of letters used to denote the name of Jesus Christ.

manuscript in Paris is from 1473. It is a translation of the *Aliabraa Argibra* into Hebrew by Mordechai Finzi, a respected scholar and mathematician from Mantova (Mantua). The following is a quote from Finzi's introduction. It gives us Maestro Dardi's name and the date of the work, and it gives interesting details such as the name and address of the scribe. (Translation from Reference [4] below.)

> Here are the rules of the six abbreviated chapters of the book *gabli almukabala* which the Maestro Dardi of Pisa wrote and explained, which are in number of 194... This book was translated from another book first written on 9 November 1344 in the Christian reckoning, and afterwards they (sic) began to write it, Yakomo d'Ierushali di Litovilana, who lives in the city of Mantua on the road of the Unicorn close to the Holy S. Byrnaby, on Saturday, 3 May 5229 [= 1469 C.E.], in which were kept many books of value in arithmetic as we can see throughout this book. I, Mordecai Finzi, began to translate it here in Mantua from the Christian [language] to Hebrew for the benefit of our people on Wednesday, 24 November 5234 [= 1473 C.E.] in the era of the creation, and I trust in God that I will not stumble.

In 1986 Barnabas Hughes (See [3] below) was asked by a book vendor to analyze the contents of a certain manuscript. It turned out to be a previously unknown copy of Dardi's work. It is now housed at the University of Arizona, Tempe. (Maestro Dardi, "Algeblle amugabala." SPEC C- 3570. Rare Books and Manuscripts, Arizona State University Library.) Images from it follow. The manuscript is in Italian. Its remarkable feature is that it has the same quote as given above, somewhat smudged, and some 800 marginal words in Hebrew. In the introduction to her transcription of Dardi's work, Raffaella Franci (See Reference 2 below) indicates that handwriting analysis suggests that Hebrew notes are in the same hand as the Paris manuscript. It could be the very manuscript that Finzi used. In any case, the *Aliabraa Argibra* has now acquired an author and a date.

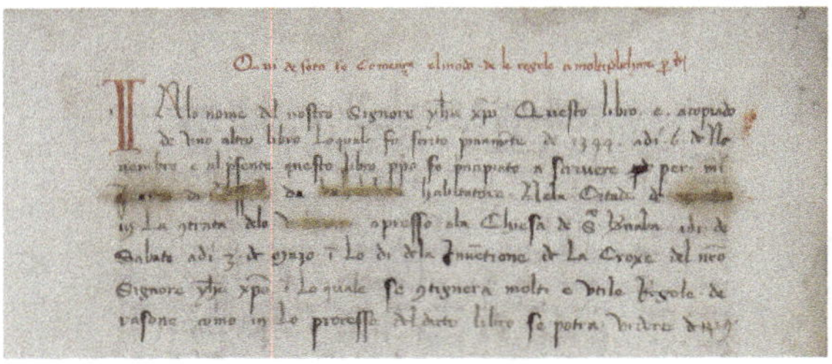

Arizona Manuscript with name and date.

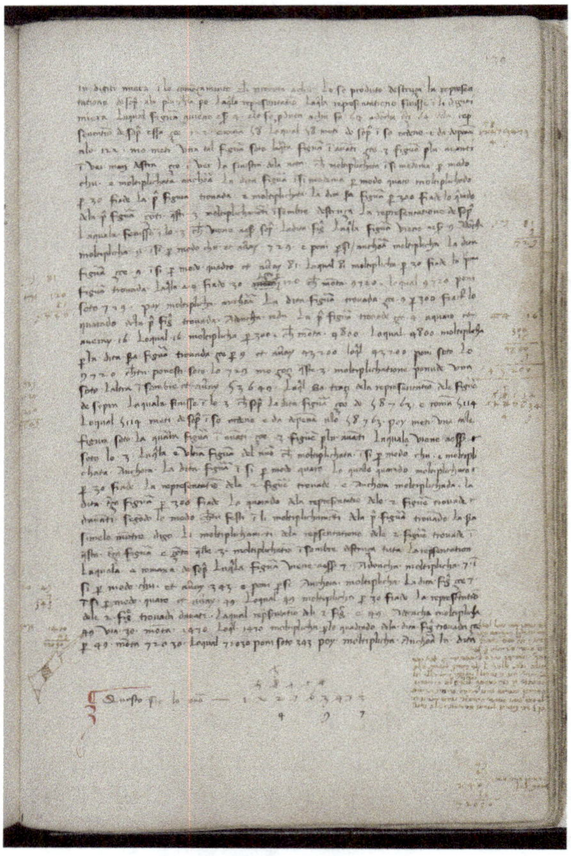

Arizona Manuscript page with Hebrew notations

Problems from the *Aliabraa Argibra*:

1. (Dardi's example for his Rule 5) Make 10 into 2 parts such that multiplying the difference that one has to the other by itself makes $20\frac{1}{4}$. I ask how much each part would be.

2. Find me 2 numbers such that each part is to the other as 2 is to 3 and multiply the first by itself and multiply that multiplication by the second number and this multiplication joined [added] to its root makes 342. I ask what each number is. (Dardi's Rule 41 addresses the case $c = ax^3 + \sqrt{bx^3}$.)

3. A man lent 100 lire to another and at the end of 3 years he receives 150 lire from him for the principal and interest compounded yearly. I ask you at what rate was the lira lent per month?

   Note: This problem involves a cubic equation not reducible to a linear or quadratic equation. It is solved correctly. The author puts the rate for one lira for one month equal to x denari. Hence after one year, one lira bears the interest of $\frac{x}{20}$ lira. (1 lira is 240 denari.) Dardi calculates capital and interest at the end of the third year, and obtains $100 + 15x + \frac{3x^2}{4} + \frac{x^3}{80} = 150$ or $x^3 + 60x^2 + 1200x = 4000$. Note that one can complete the cube to solve the problem. It becomes $100(1 + \frac{x}{20})^3 = 150$. Dardi did not approach the problem this way but solved it correctly.

References:

1. Franci, R., Toti Rigatelli, L. (1985). *Towards a history of algebra from Leonardo of Pisa to Luca Pacioli.* Janus, 72(1-3), 17–82

2. Franci R., Quarderno 26. Maestro Dardi, 2001. Aliabraa Argibra: Dal manoscritto I.VII.17 della Biblioteca Comunale di Siena, ed. Raffaella Franci. Siena: *Servizio* Editoriale dell'Universita di Siena

3. Hughes, B., *An Early 15th-Century Algebra Codex: A Description,* Historia Mathematica, 14 (1987), 167-172

4. Van Egmond, W. , *The algebra of Master Dardi of Pisa*, Historia Mathematica, 10 (1983) 399-441

Manuscript locations:

1. Paris, Bibliotheque Nationale, Ms. Hebreu 1029, ff. 194r- 234r.
2. Arizona State University, Hayden Library, Rare Books and Manuscripts; SPEC C-3570
3. Biblioteca Comunale di Siena, Manuscript I.VII. 17

Trattato di Arithmetica
By Filippo Calandri

Filippo Calandri was a member of a large family of writers and scribes. He wrote his *Arithmetica* in 1491. It breaks no new ground and it contains no algebra. Nonetheless it remains interesting for at least two reasons. The *Arithmetica* was the first illustrated math book to be produced with what was then a new technology, the printing press. It was relatively inexpensive. Now accessible to many, it became widely distributed. Its dedication is also interesting. It is to Guiliano de'Medici, son of Lorenzo de' Medici a.k.a. Lorenzo the Magnificent. Giuliano would have been 12 or 13 years old at the time, the typical age of an abbacus student in Florence. Around the same time, his father had taken in the young Michelangelo to be housed and educated with his sons. Along with the standard academic subjects grounded in classical Latin works, the Medici offspring learned the applied mercantile mathematics taught in the abbacus schools. However, the book that Guiliano would have used was not off the printing presses, but rather an exquisite hand copied book with numerous lively illustrations in the Florence miniature style.

Cover image: Pythagoras

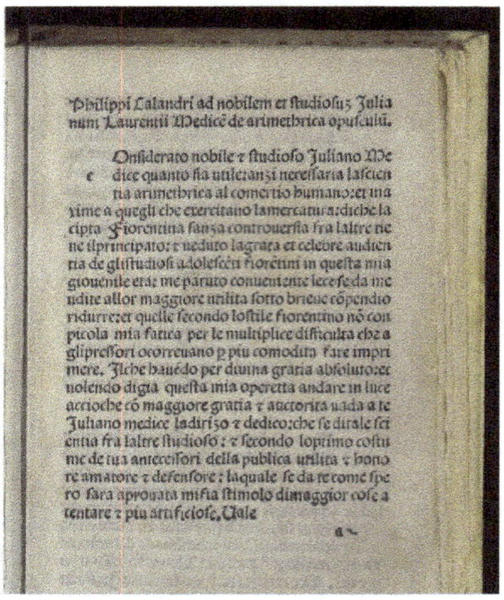
Philippi Calandri ad nobilem et studiosū Julianum Laurentii Medicē de arimethrica opusculū.

Onsiderato nobile ⁊ studioso Juliano Me
c dice quanto sia utile: anzi necessaria lascien
tia arimethrica al comertio humano: et ma
xime a quegli che exercitano lamercatura: diche la
cipta Fiorentina sanza controversia fra laltre ne
ne ilprincipato: ⁊ veduto lagrata et celebre audien
tia de glistudiosi adolescēti fiorētini in questa mia
giouentile eta: me paruto conueniente lecose da me
udite allor maggiore utilita sotto brieue cōpendio
ridurre: et quelle secondo lostile fiorentino nō con
picola mia fatica per le multiplice difficulta che a
glipressori ocorreuano p piu comodita fare impri
mere. Ilche hauēdo per diuina gratia absoluto: et
uolendo digia questa mia operetta andare in luce
acciochè cō maggiore gratia ⁊ auctorita uada a te
Juliano medice ladirizo ⁊ dedico: che se di tale sci
entia fra laltre studioso: ⁊ secondo loptimo costu
me de tua antecessori della publica utilita ⁊ hono
re amatore ⁊ defensore: laquale se da te come spe
ro sara aprouata misia stimolo dimaggior cose a
tentare ⁊ piu artificiose. Vale
a

Dedication and Preface

The cover page, the preface, and the two pages of problems that follow are from a printed edition found in the Library of Congress, https://www.loc.gov/item/48031290/. Our translation of the preface is followed by translated problems from the printed version. These are followed by a sampling of the illustrations from the De' Medici manuscript housed in the Biblioteca Riccardiana, Florence (Ricc. 2669) and printed with the permission of the Ministry of Culture.

Dedication (in Latin): Philippi Calandri to the noble and studious Julianum Laurentii Medici, this small work on arithmetic.
Preface (in Italian):
I have considered, noble and studious Juliano Medici, how useful and even necessary, the knowledge of arithmetic is to human commerce and most of all, among those who exercise mercantilism. It is said that the Florentine city is without a doubt the first among other regions. Having seen the graced and celebrated audience of the studious Florentine adolescents that were in my youth, it seemed to me necessary that the things I heard as most useful be reduced into a brief compendium. And that [it be] according to the Florentine style, with no small effort on my part, through many difficulties, the printers became more accommodated to print.
He who obtained by absolute divine grace and wanting now that my little work may go into the light, so that it goes with the most grace and authority to you, Juliano Medici, I direct and dedicate it so that

much science [will be] among your other studies. And following the best custom of your ancestors from the public service and of the good leader, lover, and defender (of that service), which if by the leader, like I hope, [it] will be approved, it will make me eager to attempt greater things and more hard work.
Vale

Four problems from two pages of the printed edition:

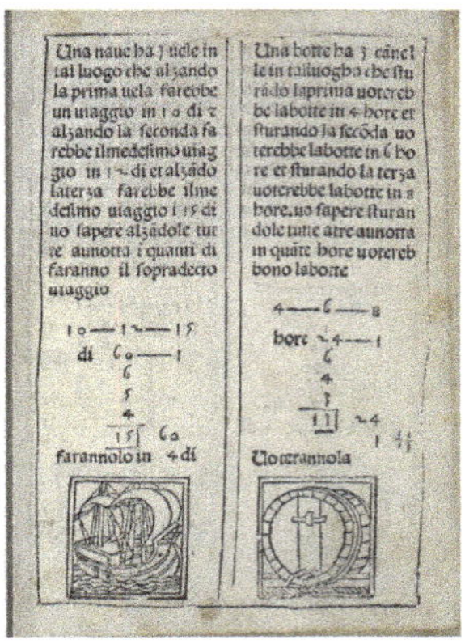

Page 1

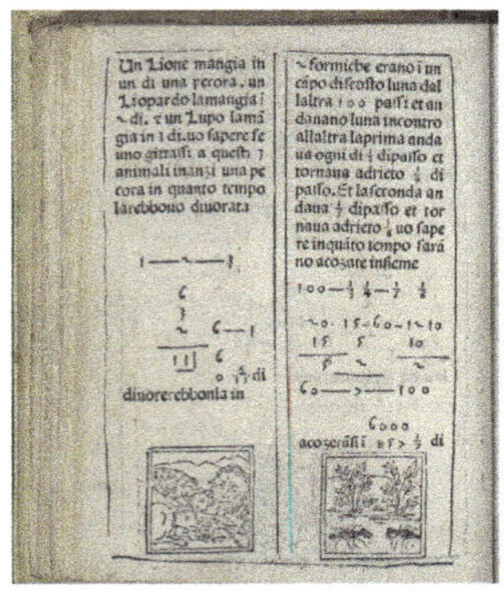

Page 2

1. The problem in left column of page 1: A ship goes from from Pisa to Genoa in five days. Another ship comes from Genoa to Pisa in 3 days. I want to know, if they each left at the same time, that which left from Pisa to go to Genoa and that which left from Genoa to go to Pisa, in how many days would they be together (in the same place)?

2. The problem in the right column of page 1: A barrel has two spigots. The first empties the barrel in 4 hours and the other empties it in 6 hours. In how many hours would the barrel be empty if they were both on?

3. The problem in the left column of page 2: A lion eats a sheep in one day, a leopard eats it in 2 days and a wolf eats it in 3 days.[1] I want to know, if one threw a sheep in front of these

[1]These are the three animals that terrified the pilgrim in the first canto of Dante's Inferno, a work that would have been very familiar to Calandri and his audience.

three animals, in how much time would it be devoured.

4. The problem in the right column of page 2: Two ants were on a promontory 100 paces away from each other and were going towards each other. The first went $\frac{1}{3}$ of a pace and turned around $\frac{1}{4}$ of a pace each day. The second went $\frac{1}{5}$ of a pace and returned $\frac{1}{6}$ of a pace each day. I want to know how much time passed until they met.

Images from the Riccardiana Manuscript (Ricc. 2669)

School Room

Trade

Shipping

Preaching

Gaming

Battle

Dining

Wine Tasting

Ars Magna by Gerolamo Cardano

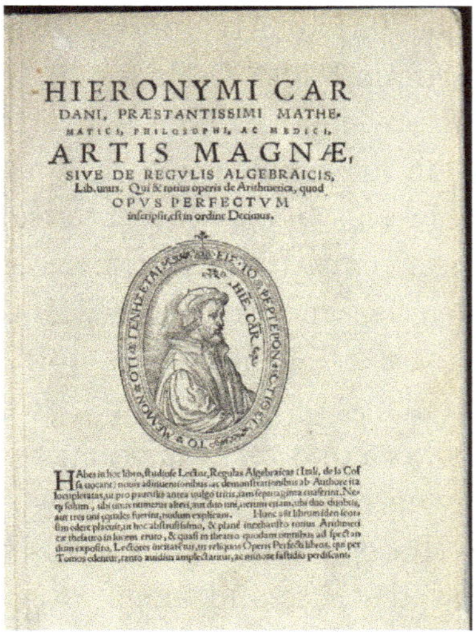
HIERONYMI CAR
DANI, PRÆSTANTISSIMI MATHE-
MATICI, PHILOSOPHI, AC MEDICI,
ARTIS MAGNÆ,
SIVE DE REGVLIS ALGEBRAICIS,
Lib. unus. Qui & totius operis de Arithmetica, quod
OPVS PERFECTVM
inscripsit, est in ordine Decimus.

The Title Page of Cardano's *Ars Magna*

A copy of Cardano's *Ars Magna* (The Great Art) is contained in the extraordinary Smith and Plimpton Collections housed at Columbia University. The MAA (Mathematical Association of America) offers an introduction to the collection and, in particular, to the *Ars Magna* on the MAA Treasures website:
https://web.archive.org/web/20240207092900/https://maa.org/press/periodicals/convergence/index-to-mathematical-treasures

V. Katz and F. Swetz introduce the work as follows:

> This is the title page from the *Ars Magna*, by Gerolamo Cardano, published in 1545. The page contains an engraving of a portrait of the author. This is the text in which the algebraic solution of cubic equations was first printed. Although the initial discovery of the solution of one type of cubic equation was due to Scipione del Ferro, and solution methods for at least three types were worked out by Niccolò Tartaglia, it was Cardano, along with his student Lodovico Ferrari, who worked out the details for thirteen

cases and then published them in this book. The book also contains Ferrari's basic method of solution for quartic equations, as well as much else.

The title page may be translated as follows:
'Book one of The Great Art, or the Rules of Algebra, by Gerolamo Cardano, most distinguished mathematician, philosopher, and physician, which is the tenth in order of the whole work on arithmetic, which is titled the perfect work. You have in this book, diligent reader, the rules of algebra (in Italian, called rules of the coss), so abounding in new discoveries and demonstrations by the author, more than seventy, that earlier works now amount to little (or, in the vernacular, are washed out). It disentangles the knots not only where one term is equal to another or two to one, but also where two are equal to two or three to one. It is a pleasure, therefore, to publish this book separately so that, this most abstruse and unsurpassed treasury of all of arithmetic being brought to light, and as, in a theater exposed to the sight of all, its readers may be encouraged and will all the more readily embrace and with less aversion study thoroughly the remaining books of this perfect work, which will be published volume by volume.'
(Translation adapted from that of T. Richard Witmer, *The Great Art*, Cambridge: MIT Press, 1968.)

The following dedication and problems are also from Witmer's translation. We note that the original was in Latin, and, although it filled with abbreviations, remains rhetorical. Some of the notation in the translation is modern rather than original.

Cardano's Dedication

Girolamo Cardano, Physician, to the most erudite *Andreas Osiander*,[1] greetings:

[1]Andreas Osiander (1498-1552) of Nurenberg was one of the minor leaders of the Reformation. He is generally credited with authorship of the anonymous preface to Copernicus' De Revolutionibus.

I have considered nothing so deeply, learned Andreas, as the names of those who, by their writings, deserve to be commended to posterity. I have especially asked whether they combine liberal learning with erudition. Hence, since I know that you have a far from mediocre knowledge not only of Hebrew, Greek, and Latin letters but also of mathematics and since, moreover, [you are] the most humane man I have ever encountered, it seemed to me that this, my book, could be better dedicated to no one than you by whom it may be corrected (if my pen has gone further than the power of my mind), read with pleasure, understood, and, indeed, recommended authoritatively. Unless I am mistaken, others will follow this example and dedicate their works [to you], whatever branch of learning they pursue. Accept this, therefore, as a perpetual testimonial of my regard for you, of your assistance to me, and, above all, of your distinguished scholarship. And, although you may well be such that your merits are known to all men, yet just as Alexander and Caesar wished to have their most notable deeds inscribed in the chronicles of others and as Plato, who preserved such marvels in his own writings, still desired to be praised in the writings of others, so I hope that this, my offering to you, will not be displeasing, whatever its merit, because in such matters there is a certain fortune which controls and the better perish while the poorer survive. And whatever your judgment may be in matters of this sort, it nevertheless is certain to me that it is my duty to repay my debts. I hope also that, by this clear example, there may be known my affection for all who have that same candor of spirit for which you are recognized among the scholars of our time. But perhaps a better occasion [for this] will be given; if not, I still do not wish this one, such as it is, to be lost to me. Farewell. Pavia, the 5th of the Ides of January, 1545.

We present two examples from the *Ars Magna*, both of which were milestones in the history of algebra. In the first, we find the first appearance of a complex number. In the second, we have a rule (correct!) for solving a cubic. Before the presenting this rule, Cardano gave a proof that its steps worked. The proof was geometric and is not given here.

1. If someone says to you, divide 10 into two parts, one of which multiplied into the other shall produce 30 or 40, it is evident that this case or question is impossible. Nevertheless, we shall solve it in this fashion.[2]

   Let us divide 10 into equal parts and 5 will be its half. Multiplied by itself, this yields 25. From 25 subtract the product itself, that is 40, which as I taught you . . . leaves a remainder m: 15. The root of this added and then subtracted from 5 gives the parts which multiplied together will produce 40. These, therefore, are 5 p: R m: 15 and 5 m: R m: 15. (i.e. $5+\sqrt{-15}$ and $5-\sqrt{15}$).

2. (A correct rule for solving a cubic of the form $x^3 + bx = c$.) Cube one-third the coefficient of x; add to it the square of one-half the constant of the equation; and take the square root of the whole. You will duplicate this, and to one of the two you add one-half the number you have already squared and from the other you subtract one-half the same. You will then have a binomium and its apotome*. Then, subtracting the cube root of the apotome from the cube root of the binomium, the remainder [or] that which is left is the value of x. For example, $x^3 + 6x = 20$. Cube 2, one-third of 6, making 8; square 10, one-half the constant; 100 results. Add 100 and 8, making 108, the square root of which is $\sqrt{108}$. This you will duplicate: to one add 10, one-half the constant, and from the other subtract the same. Thus you will obtain the *binomium* $\sqrt{108}+10$ and its *apotome* $\sqrt{(108)}-10$. Take the cube roots of these. Subtract [the cube root of the] *apotome*[3] from that of the *binomium* and you will have the value of x:

$$\sqrt[3]{\sqrt{108}+10}-\sqrt[3]{\sqrt{108}-10}.$$

[2] Cardano uses 40. He uses the abbreviation m: for minus, p: for plus and R for root.

[3] An *apotome* is a number of the form $a-\sqrt{b}$ where a and b are rational numbers and similarly a *binomium* is a number of the form $a+\sqrt{b}$.

A Postscript: Finding the correct solution to the cubic equation was a milestone in the history of algebra but it precipitated a famous contentious battle for intellectual property rights. The principal players were Ferrari, Cardano, and Tartaglia. In her article, "Niccolo Tartaglia's poetic solution to the cubic equation", Arielle Saiber gives a thoroughly humanistic account of the events surrounding Cardano's publication of the solution to the cubic. (The role of math competitions may be a surprise.) The article gives us a bonus, a copy and translation of Tartaglia's poem in which he gives a rule for solving a cubic. Here's the poem, *Quando Chel Cubo* or *When the cube*, with translation and algebra by Veronica Gavagna.

1.	Quando chel cubo con le cose appresso Se agguaglia à qualche numero discreto	1.	When the cube and things together are equal to some number,	1.	$(x^3 + px = q)$
2.	Trouan dui altri differenti in esso	2.	find two other numbers differing in this one	2.	$(u - v = q)$
3.	Dapoi terrai questo per consueto Che'llor produtto sempre sia eguale Al terzo cubo delle cose neto,	3.	that their product should always be equal exactly to the cube of a third of things.	3.	$(uv = (p/3)^3)$
4.	El residuo poi suo generale Delli lor lati cubi ben sottratti Varra la tua cosa principale.	4.	The whole remainder of their cube roots subtracted will be equal to your principal thing.	4.	$(x = \sqrt[3]{u} - \sqrt[3]{v})$
5.	In el secondo de cotesti atti Quando che'l cubo restasse lui solo	5.	In the second of these acts, when the cube remains alone,	5.	$(x^3 = px + q)$
6.	Tu osseruarai quest'altri contratti, Del numer farai due tal part'à uolo	6.	you will observe these other agreements: you will at once divide the number into two parts	6.	$(u + v = q)$
7.	Che l'una in l'altra si produca schietto El terzo cubo delle cose in stolo	7.	so that the one times the other produces clearly the cube of the third thing exactly.	7.	$(uv = (p/3)^3)$
8.	Delle qual poi, per commun precetto Torrai li lati cubi insieme gionti Et cotal somma sara il tuo concetto.	8.	Then of these two parts, as a habitual rule, you will take the cube roots added together, and this sum will be your thought.	8.	$(x = \sqrt[3]{u} + \sqrt[3]{v})$
9.	El terzo poi de questi nostri conti Se solue col secondo se ben guardi Che per natura son quasi congionti.	9.	The third of these calculations of ours is solved with the second if you take good care. as in their nature they are almost matched.	9.	$(x^3 + q = px)$
10.	Questi trouai, & non con passi tardi Nel mille cinquecentè, quatro e trenta Con fondamenti ben sald'è gagliardi Nella Citta dal mar'intorno centa.	10.	With quick steps and light feet these solutions I found, in one thousand five hundred, thirty and four, my foundations quite sure and certainly sound in the watery city surrounded by shore.		

Tartaglia's Poem

Resources:

1. Cardano, Gerolamo, *Rules of Algebra, 1545 (Artis Magnae, sive de regulis algebraicis)*, Trans. Richard Widmer, Dover 2007

2. Cardano, Girolamo, 1501-1576 (in the original Latin): Hieronymi Cardani, præstantissimi mathematici, philosophi, ac medici Artis magnæ, sive, De regulis algebraicis .. Smithsonian Libraries and Archives - Digital Library

3. Saiber, Arielle, (2014). "Niccolo Tartaglia's poetic solution to the cubic equation," Journal of Mathematics and the Arts, 8(1–2), 68–77. https://doi.org/10.1080/17513472.2014.933552

www.ingramcontent.com/pod-product-compliance
Lightning Source LLC
LaVergne TN
LVHW082349200926
848481LV00012B/3507